## NATURE AS LANDSCAPE
### Dwelling and Understanding

Human interaction with the natural world has created serious environmental problems. In *Nature as Landscape* Kraft von Maltzahn examines the "unnatural" relationship that has developed between human beings and nature. Drawing on evidence from philosophy and the history of science, he argues that humans have become estranged from nature and shows how this estrangement has evolved.

Von Maltzahn focuses on how we experience aspects of nature in terms of their outer appearance, such as landscape, and contends that the naturalistic scientific tradition has taught us to divorce ourselves from the natural world, to become impartial observers rather than participants. He examines the nature of the human life-world and describes the process of self-deception that has led to the contemporary dismissal of that life-world as merely subjective. Drawing on phenomenology, semiotics, visual thinking, gestalt psychology, and Polanyi's arguments about tacit knowing, he offers an alternative way of perceiving the natural world that would reunite humans and nature.

Given the current state of the global environment, it is crucial that the debate on the relationship of human beings and nature take place on many levels. Given the failure of conventional approaches to conservation, developing alternative ways of understanding human relations has become critically important.

KRAFT E. VON MALTZAHN is professor emeritus of biology, Dalhousie University and University of King's College.

K.H. Steib, *Nach der Ernte*, 1976 (Kraft von Maltzahn, Halifax, Nova Scotia)

# Nature as Landscape

## Dwelling and Understanding

KRAFT E. VON MALTZAHN

McGill-Queen's University Press
Montreal & Kingston • London • Buffalo

© McGill-Queen's University Press 1994
ISBN 0-7735-1233-0

Legal deposit third quarter 1994
Bibliothèque nationale du Québec

Printed in Canada on acid-free paper

This book has been published with the help of a grant from the Social
Science Federation of Canada, using funds provided by the Social
Sciences and Humanities Research Council of Canada. Funds have also
been received from Dalhousie University and from the Canada
Council through its block grant program.

---

**Canadian Cataloguing in Publication Data**

Von Maltzahn, Kraft E. (Kraft Eberhard), 1925–
Nature as landscape: dwelling and understanding
Includes bibliographical references and index.
ISBN 0-7735-1233-0
1. Human ecology – Philosophy. I. Title.
GF90.V65 1994    304.2    C94-900343-3

---

This book was typeset by Typo Litho Composition Inc. in
11/13 Adobe Garamond.

# Contents

# Figures

# Preface

The thesis presented in this work arose out of a conflict concerning our own position in relation to nature. We were taught to be impartial observers of nature in order to gain objective knowledge of the natural world. But we became aware that we cannot be impartial observers of nature, because all our experience is necessarily intentional. Knowing is not an exercise in impartiality, but knowing and being are intricately linked in our everyday life-world.

In my attempt to resolve some of the conflict between the scientific attitude and a view of science as only one aspect of the life-world, I have been helped by students and masters alike. Professor A. Frey-Wyssling of the Eidgenössische Technische Hochschule in Zürich, Switzerland, shared his wide range of skilful knowledge with his students, introducing us not only to alpine plants and the landscapes containing them, but also to the submicroscopic morphology of protoplasm. Professor E.W. Sinnott of Yale University in Connecticut encouraged me during my doctoral studies to view even the most specialized problem in the broadest context.

I thank my colleagues Cheryl Knight and Dr Pat Lane for helping me in the early stages of the manuscript, and especially my wife Anne for clarifying my written expression. Pat Dixon patiently and cheerfully typed each version of my manuscript. Dr Warwick Kimmins, dean of science at Dalhousie University, enabled me to obtain financial assistance for publication of my work. I am grateful to my publishers, especially Peter Blaney, for his understanding and support, and to Carlotta Lemieux for her perceptive copy editing.

Figures 4 and 5 are reproduced by courtesy of the trustees, the National Gallery, London. Figure 6 is reproduced by permission of the Kunstmuseum in Berne and figure 3 by courtesy of S. Haase. The frontispiece, *After the Harvest* by K.H. Steib, illustrates a principal theme of the book, and I thank the artist for permission to use his work. I also thank those many students who discussed with me aspects presented in this book; without them I could not have written it.

*Nature as Landscape*

# Introduction

One is immensely alone among trees which flower and among brooks
which pass by. One feels not so abandoned by far, when alone with
a dead human body, as when alone with trees. As mysterious as death
may be, far more hidden is the life which is not our life, which does
not partake in us.

Rainer Maria Rilke

In his essay "Worpswede," Rainer Maria Rilke describes our relationship
to the concrete aspects of nature as we encounter them in our everyday
life, the "trees which flower" and the "brooks which pass by."[1] What he
describes is our complete estrangement from nature, an estrangement
that frightens us once we become aware of it. But it is not merely fright
that wells up in us. There is also puzzlement, for we are born and raised
within a particular life-space on earth, surrounded not only by nature
with its characteristic features – the contour of the land and the life-
forms of the plants, singly and in their whole composition – but also by
the works of human beings. Through our day-to-day experience of
everything that surrounds us, we ought to obtain a sense of identity with
nature, a sense of home. Instead, we realize that we have no home, and
we inquire how this state of homelessness has come about.

We have learned to consider the world as the object of knowledge.
When I say "we," I speak of all contemporaries, including myself,
brought up in the naturalistic scientific tradition. According to naturalistic cognition, nature is composed of bodily things such as rocks, plants,
and animals – natural bodily things. We assume that we gain cognition
of the objects of nature on the basis of the unalterableness of these objects and our direct access to them. We become conscious of what *is* by
turning to the objects as they are given to us directly. In the naturalistic
scientific tradition, we wish to attain the knowledge of nature in the positive sense. In other words, this knowledge must be independent of the
individual person. The world it represents must exist irrespective of

whether you and I are present. This requires that we adopt the perspective of the impartial observer, the attitude of the spectator in relation to a spectacle, and that we do not become involved.

Nature as external reality consists of extended homogeneous space that contains beings, such as a granite rock or a grasshopper. In the naturalistic scientific tradition, it is our task as observers to study beings in our quest to know the external world. Space is conceived as homogeneous and neutral. Because beings contained in a space do not exhibit any relationship to that space, they are readily replaceable. In this regard, we can convert beings into numbers, but changes in numbers are expressible as a function of a homogeneous and continuous time scale. In this way, nature is explained in terms of mathematical principles.

Nature in terms of the "one objective world" is composed of material bodily elements that must be laid bare. The scientist explains the phenomena of nature in terms of the underlying elementary natural forces. One is capable of knowing the truth about nature in terms of the language of science based on physical reason. What takes place in nature is due to the forces of nature. The scientific study of nature therefore concerns itself with "that which is possible by way of the state and the forces of bodily things."[2] We construct the laws of nature in thought and consider this construction to be authoritative because it allows us to make predictions about natural events in advance of experience. Predictability presents us with certitude in the face of chance, and the certitude enhances the effectiveness of our decisions.

There is not only the world of science, there is also the world of our everyday experience – our "life-world." The biologist Jakob von Uexküll spoke of these two worlds and how they are related, one to the other, when he recounted in his autobiography the story of a man who began to think of his own shadow as an independent person. Initially, the shadow was the man's servant, but it gradually became the master, while the man in turn became the servant of his own shadow. Why did Uexküll begin his autobiography by telling of this strange relationship between a man and his shadow which ultimately results in the man behaving absurdly? Uexküll himself gives us the answer: "It is easy to draw the conclusion from this story. Most natural scientists have unfortunately not done so and have instead made the shadow the master in their world view."[3]

Uexküll thus refers to the positivists, who view objects as unchangeable. They speak of sensory illusion if one and the same object is perceived differently by different individuals. How, then, does Uexküll justify comparing the positivists with the shadow that becomes the owner's master? The positivists' one objective world was intended to serve us for our convenience in our everyday life, but instead our everyday life-world has become subservient to the one objective world. But are objects really unchangeable, as the naturalistic positivist claims? Let us look first at the surroundings of different kinds of animals, as Uexküll did. Each kind of animal shapes its own surrounding world. Whatever enters the surrounding space of an animal is given meaning by that animal according to its life-form. Animals belonging to different life-forms (for instance, predators and grazers) have totally different surroundings. Accordingly, an object, which is unchangeable within the objective world, may take on very different meanings, depending on the surrounding world it enters.[4]

Each animal has a surrounding world that is characteristic and fixed for its kind. You and I and all human beings have surroundings too, but they are not fixed. The meaning we give to things and beings as they enter our world is determined by our intentions. We confer meaning on the world by what Edmund Husserl calls "intentionality."[5] He is referring to consciousness and he understands consciousness to be an intentional experience. Intentionality implies that there is someone who intends and something that is intended. Consciousness points intentionally beyond itself to what we are conscious of.

We are capable of building very different worlds. The objective world, as we construct it in science, is merely one of these worlds. It is a world that we cannot experience as such, because it is a world we have constructed in thought. In contrast to this constructed world, there is the world as it may appear before each of us, the phenomenal world of direct experience. This is the life-world, our world of everyday experience, the reality we live. When we speak of life-world, we need to refer to our body as the instrument of our experience – not, however, as we know it from the outside but as we perceive it from inside.

I stated above that we are surrounded by nature as well as by human constructs. We refer to all this, taken together as it relates to us, as "environment." Although environment is literally that which environs, that which surrounds, today we understand it to mean principally the conditions affecting the life of human beings. In the study of the environment, we become overwhelmed by our naturalistic scientific tradition and view our environment as an object. The world of objects is the concern of science. Since the study of the environment investigates the environment objectively, it clearly must do so in the tradition of science.

Let us return, though, to the literal meaning of "environment" as that which environs, that which surrounds. It becomes evident that a human environment is only one of many different forms of environment. Organism and environment can be defined only on the basis of the relations that exist between them. It is a configuration. We give meaning to a configuration, and in this manner experience arises. For example, primarily (though not exclusively) through the act of seeing, we shape aspects of nature into landscape. We experience aspects of nature in terms of their outer appearance, such as landscape. The outer manner of appearing may reveal to us the inner nature of the aspect. This relationship is called "physiognomy." The inner character is a form, and reading a physiognomy is a skill we have to learn. But our understanding of a physiognomy is preconstituted in our life-world in the form of familiar patterns and meanings.

We begin to recognize that we need to interpret the world. Interpretation implies the act of understanding, but understanding necessitates that we enter into a relationship with those aspects we wish to understand. We can place ourselves opposite the landscape and separate from it. We can also incorporate ourselves into the landscape and participate in it. In this case, our acts in relation to the space of the landscape may incorporate our whole existence. To be merely an impartial observer will not do. Understanding implies dwelling; in order to understand, we must learn to inhabit the world.

In the words of the Spanish thinker Ortega y Gasset, "Culture presents us with objects already purified."[6] The objects are purified by abstract thinking. I suggest, then, that we have come upon our sense of being "immensely alone among trees" because our consciousness has already separated itself from the trees and has purified them as mere objects. Nature is conceived as the extended bodily world, opposite us and separate from us.

The appraisal of our life-world as being merely subjective in contrast to the ideal of the knowledge of the one objective world forms an important part of our naturalistic-positivistic tradition. Does this valuation not contribute significantly to the human-environmental issues afflicting us? In the following pages, I shall briefly characterize features of this traditon and explore their impact on the human-to-nature configuration; and I shall look at alternative structures of being and thinking that are more appropriate for us in building environments.

### CHAPTER ONE

# The Objective Interpretation of Nature

## THE NATURALISTIC SCIENTIFIC VIEW OF NATURE

Nature is that which brings itself forth, and it exists as it has brought itself forth whether we are present or not. Nature is composed of natural bodily things; and natural bodily things, in contrast to things made by people, are shaped by the forces of nature. Natural cognition gives us direct access to the items that constitute nature, and the knowledge of these items permits us to build up an inventory of nature.

The Swiss physician Conrad Gesner was one of the early naturalists who attempted to make such an inventory. In June 1541 he wrote to his friend Jakob Vogel, describing a trip he had taken into the mountains. Gesner stated that he intended to undertake these excursions a number of times every year in order to gain knowledge of the plants and enjoyment of the mind. His letter consists to a large extent of explanatory descriptions of his observations of the natural phenomena in these mountainous regions. He accounts, for example, for the favourable growth of forests at higher altitudes in terms of the nutrient contents of the water in the ground.

A principal reason for Gesner's trip was a desire to become acquainted with the plants. What he means by "knowing" we can learn from his *Thierbuch* (Book of Animals),[1] the work by which he is best known and for which he is considered a founder of the field of zoology. The work consists of four volumes in which the animals are arranged alphabetically according to the first letter of their Latin names. Gesner lists first the names given to each animal in various languages. Next, he informs the reader about the native region of the animal and describes its various organs. He also discusses the animal's behaviour and lists the possible uses the animal might have when tamed or at the hunt, as well as its use for food and medicine. Finally, he comments on the animal in mythology and its symbolism. Gesner attempted, too, to assemble the best illustrations available at the time, such as Albrecht Dürer's etching of a rhinoceros. An identity emerges between what is observed of some object of nature in its outer appearance, its pictorial representation, and what is written about it.[2] Because, for Gesner, animals are objects, he arranges them in encyclopedic order in a completely isolated manner so that their use to man may be readily recognized.

The rational structure of our knowledge of nature requires that we form concepts of objects with common properties and concentrate upon their similarities and therefore abstract their essential properties. The concept "tree" may serve as example. There are many different kinds of tree, but the concept of "tree" concentrates on those features that all the different trees have in common; namely, they are natural objects, they are woody plants, and they usually form but a single trunk from which arise branches of various orders; they also have a long life expectancy and normally become much larger than most other kinds of plant.

When we wish to discover the diverse objects nature contains, as Gesner undertook to do, and set out to reveal the order underlying the diversity of objects, we must be able to identify them. This requires naming them. We want to establish a one-to-one relationship between a particular kind of being and its name, setting it apart for identification. The great naturalist Linnaeus considered that plants and animals exist in an immutable order and that this order is expressed by the degree of similarity and difference between various kinds of plants and animals. This order can be discovered through comparison based on the selection of appropriate characters. The structural features chosen in order to establish identity or difference between individuals are the differentiating characters.

In his system of classification, Linnaeus chose to give the flower a privileged position and, within the flower, the reproductive organs. He thus established a sexual system (*systema sexuale*) based on the number of stamens and pistils of the flower, and their shape and size in relation to each other and to the shapes, sizes, and number of the other floral organs.

This is illustrated by a dispute between Johann Jacob Dillenius, professor of botany at Oxford University, and Linnaeus, who was then visiting Oxford. Dillenius had described the flowers of a *Chenopodium* as having three stamens, but Linnaeus maintained that this was not so and he demonstrated the fact to Dillenius: "I opened the flower and showed him that it had only one. 'No doubt it's an abnormal specimen,' he [Dillenius] said. We opened several more, and they were all the same. We passed on to several other genera, and all tallied with my description of them. Dillenius was amazed."[3]

Knowledge about plants, as one kind of object of nature, is sought through painstaking observation and the selection of appropriate characters, such as the number of stamens. The similarities and dissimilarities between the objects are established on the basis of comparisons of the characters selected. The systematic order is founded on the degrees of semblance which the comparisons have yielded. Linnaeus called the characters taken together in their true associations for the establishment of identity or difference, as well as their level of difference, "natural characters." The system of order based on natural characters is the "natural system" of classification. Selection of structural features that are not indicative of true associations are "artificial characters"; a system of classification founded on such characters represents an "artificial system." In Linnaeus's view, a natural system of classification was virtually unobtainable. He therefore developed an artificial system based on the sex organs of the flowering plants.

Organisms that are very similar to one another and are thus considered to be of the same kind are placed together into the same "species." A higher, more inclusive category into which a number of similar species are placed is the "genus." Linnaeus gave each kind of organism two names, that of the genus followed by the name of its particular species. His system is therefore also known as the system of "binomial nomenclature." An example is the scientific name of the eastern flowering dogwood: *Cornus florida* L., where *Cornus* represents the genus, *florida* the species, and L. stands for Linnaeus, the author who first described the species. A plant is thus characterized by its name (which often is a translation describing some of its characteristics). The correspondence of thing or being and word or name expresses also the position of each kind within the whole immutable hierarchical order of different kinds of beings. The approach to the phenomena of nature in the form of "natural history" was intended to give us knowledge about "what is," in the form of an inventory of nature. Linnaeus believed his system of order was not the final system, because the systematist's aim was to find a natural system of classification while his own had to remain an artificial one.

About a century after Linnaeus, the great British naturalist Charles Darwin wrote in his autobiography: "In September 1858 I set to work by the strong advice of Lyell and Hooker to prepare a volume on the transmutation of species, but was often interrupted by ill health ... I abstracted the MS begun on a much larger scale in 1856, and completed the volume on the same reduced scale. It cost me thirteen months and ten days hard labour. It was published under the title of the *Origin of Species* in November 1859 ... It is no doubt the chief work of my life."[4] What is the meaning of the title of Darwin's work? The question of the origin of species did not arise for Linnaeus because he believed that the whole order of nature, once created, remained unchanged. Darwin collected evidence that new species arise and that they have done so throughout the history of the organic world (which has been a very long one), and the question of their historical origins as well as the physical reason underlying their origin became a central one for him. What are his theses in this regard? What effect do they have on the attempt to comprehend the systematic order among the kinds of beings comprising organic nature?

According to Darwin, organisms vary from one another by their nature. Because there is only very limited space available for the unfolding life of organisms, and because individuals increase in number at very high rates, a struggle for life exists between them. Only the fittest individuals will be able to survive in this struggle. The characters that make the surviving individuals most fit will be inherited, and their offspring also will have a better chance of surviving. The variation most useful in this struggle will thus be preserved – a process known as "natural selection." Since individuals belonging to the same species have approximately the same space requirements, the struggle between them is the most severe. The creature that survives is improved in relation to its condition of life, and this results in a gradual advance in organic form. This leads, furthermore, to a divergence of characters so that a particular kind of space can be occupied by a greater number of different forms. According to Darwin, the small differences between varieties within one and the same species will, as a result of these processes, give rise in time to the larger differences between species. Darwin held that this combination of events accounted for the contemporary hierarchy of organismic forms. The underlying historical aspects of the change of form become the point of focus.

Linnaeus had discovered the constant order in diversity by comparing selected features. For him the central question was, What is similar and what is dissimilar, and what is the level of dissimilarity? He knew that he would not be able to find a truly natural system of classification, but he hoped that in the hierarchical system of constant order he would not deviate too far from it. For Darwin, that constant order dissolved into a struggle for survival between individuals, especially related ones, a struggle that brought about a continuing change of the organic world. The historical momentum thereby became the overriding one, and the principal issue was to establish pedigrees by asking, What is primitive and what is advanced? Thus, the taxonomic system emerged, founded on genealogy – the idea of finding natural systems of classification that reflect true relationship of descent.

But what features of the organisms is one to take into account? Since correlations of similarities of various characters are used in the construction of natural groups, one must give higher importance to those features that are conservative – the ones that change least and are therefore relatively stable characters. Floral features are such characters and are therefore especially suitable for structuring more natural systems of classification of the flowering plants, based on descent. Linnaeus's emphasis on the floral features in his construction of the systematic order of that kingdom of organisms was well founded.

It must be emphasized, too, that a taxonomic system based on genealogy has a high degree of predictive value. We noted above that in the search for natural groups, correlations of similarities of characters are used. Thus, when one has found such a group, it is highly probable that if some character not used in structuring the group is found in one member, it will also be found in the other members of that group. For example, the Caryophyllales represent an order of plants established principally on the basis of similarities in floral characteristics and features of the plants' embryos. In most plants, the pigments responsible for the red to yellow colours of the floral organs are chemical compounds called anthocyanins, but in almost all the families of Caryophyllales these colours are produced by a group of pigments called betalains. When we know that one member of this order contains this kind of pigment, we can predict that another member most likely will, too.

The orders that Linnaeus and Darwin elucidated are based on naturalistic procedures. They wished to account for "what is" in terms of the beings themselves, isolated as natural objects from the ground into which they are embedded and hence, as objects, separate from and opposite the observer. In order to arrive at a classificatory order, Linnaeus not only had to isolate the plants from their settings but he also had to select characters from these plants, such as the sex organs of their flowers. The classificatory order had to be founded on comparative grounds, establishing degrees of resemblances and differences of these features.

But there is another order of nature, more directly linked with our experience of our surroundings. This order is landscape. Landscape relates to where we are located on the earth. At the turn of the eighteenth to the nineteenth century two explorers, Alexander von Humboldt and Aimé Bonpland, were already concerning themselves with the spatial geographical order, principally with that of plant life-forms. In 1805 they published their *Essai sur la géographie des plantes*. The concept of the geography of plants applied to changes in vegetational composition, both with changing altitude and changing latitude. In this inquiry into spatial order, Humboldt wished to know "what is" in terms of the spatial compositions. Even though he also based his investigations on the naturalistic approach, he sought to discover the spatial order underlying the unity of nature. Such an approach requires that we do not isolate objects from the ground but that we experience them as an aspect of figure-to-ground relations in the form of landscape, a composition in which the plants form a significant part.

Humboldt early developed a profound interest in the nature of landscapes, recognizing plant forms as an integral part of landscape. He therefore had to concern himself with the important question of the relationship between the kind of landscape, its possible determination by the dominant class of plant life-form, and the geographical location. His experience of different landscapes and his observations of their vegetation made it evident to him that there are only a few basic plant forms. In his analysis of the diversity of landscapes of the tropical regions of South America, he distinguished seventeen classes of dominant plant forms that profoundly influence the characteristics of the landscape; these included the banana form, the palm form, and the vine form.

Humboldt emphasized that these types of plant form are not based on the characteristics of the reproductive organs that may be indicators of lineage – the characteristics used by Linnaeus to establish the systematic order of the flowering plants; rather, they are based on the so-called vegetative organs that make up the bulk of the plant as it participates in shaping the visible form of the landscape. Humboldt accordingly explored the spatial and geographical order with regard to these life-forms.[5]

As we now know, genealogically unrelated plants often have a similar appearance. Thus, principal types of plant forms may repeat themselves in very distant geographical areas that have similar conditions. In high mountainous and relatively bleak plateaus in the tropics, conspicuous elements of the vegetation are plants composed of a stem, which may be several metres tall, bearing leaves near and around its tip in a rosettelike manner; from the centre of this rosette arises a long, straight, candlelike cluster of flowers. In Bolivia in South America, these plants are *Espeletia grandiflora*; in northern Ethiopia in Africa, they are *Rhynchopetalum montanum*. Features of the bulk of the plant show trends towards convergence, indicating that the plants are becoming similar in similar conditions of climate and soil.

Depending on their point of view – their perspective – students of the natural world may discover very different orders underlying the phenomena of nature – for example, a hierarchical systematic order, a genealogical order, and, as Humboldt discovered, a spatial order. These approaches to the study of nature in terms of natural history are all intended to give us knowledge about "what is" by making an inventory of nature and enabling us to identify the objects of nature singly or compositely.

## THE THING-CONCEPT VERSUS THE CONCEPT OF LAW

When forming an inventory of nature, we thus initially want to know "what is." A science is therefore principally concerned, in its early stages of development, with the thing-concept – what kinds of things there are. The philosopher Martin Heidegger has distinguished three traditional ways of looking at things: (1) The thing is the bearer of its characteristic traits. A block of granite is extended, shapeless, rough, hard, and heavy. It is also coloured and partly dull and partly shiny. (2) The thing is what is perceptible by means of sensations; it is nothing but the unity of the manifold of what is given in the senses. (3) The thing is shaped matter.[6]

Our reason needs features that differentiate one kind of thing from another. We conceive a certain number of objects as having common properties by abstracting from their differences. In this way we can reflect upon their similarities and gain a general idea of such-and-such a kind of object. The concept designating the kind of object is the idea representing the essential properties of the object under consideration.

This method of proceeding was necessary, for instance, for Linnaeus to develop a systematic order of the organic world. It involves both inspection and description. This implies that one not only needs to observe the object and to represent it pictorially but that one must also describe it by means of language.

Accordingly, we conceive of nature as being demarcated by naturalism, positivism, and objectivism. The objects of nature must be perceived and represented in their outer correctness, and this must be done from one fixed point of view. This is known as "naturalism." For instance, in his etching of a rhinoceros, Albrecht Dürer represented the animal in pictorial form in its visible, tactile, and measurable aspects. Visible external reality consists of corporeal things, contained in space, which are composed of material substances and which exhibit colour according to their surface features.

Let us consider also the picture of a landscape. According to naturalism, the space of the landscape within the picture is created by means of central perspective, air perspective, colour perspective, and the overlapping of the objects contained within the picture. Natural objects or objects produced by humans are represented on the plane surface of the picture as they appear to the eye: the actual objects of the surrounding space must be transformed into elements of the picture. This is achieved by means of linear perspective as well as by modelling with light and shade. Linear perspective is given through the relationship of the diagonal lines running directly into depth and the lines running parallel to the surface. Air perspective implies an increase in brightness, a decrease in the contrasts between brightness and darkness, and a decrease in the sharpness of detail, in each case from foreground to background.[7]

Central perspective projection implies that the observer's eye is fixed in one particular position – that the scene is located in front of the observer and at a fixed distance and direction away. Because both the observer's position and the position of what is being observed are stationary, the dimension of time is not an element. The attention of the observer is controlled by the vanishing point of the perspective projection. It is this that links the observer with the scene of the picture. At the same time, placing the pictorial space opposite the observer makes it into a container, holding objects that we may isolate at will from their ground and observe in their outer appearance from one fixed point of view.

Naturalism strives to represent the factual stable truth of the outer reality. The individual must submit to the authority of the bodily objects.

The nineteenth-century painter Gustave Courbet, for example, believed that painting must represent the bodily objects the artist can see in front of him, towards which he can reach and which he can touch. Courbet and his contemporaries believed that their primary task was to imitate aspects of the world pictorially and to elucidate their material nature. The artist forms the centre and acts the role of stationary consciousness. The world is the object of knowledge and representation, and the painter must record the objective things in terms of their correct shapes and colours. Even the artist as writer must examine the subject matter with the same keen observation, analysis, and objectiveness that the naturalist painter brings to the phenomena of nature. The naturalist writer Wilhelm Bölsche recommends: "[These people's] passions, their relations to external circumstances, the whole play of their thoughts follow certain laws which the researcher has discovered and which the poet must observe in his free experiment as carefully as the chemist, if he wishes to make something rational and not some worthless mixture. He must calculate the forces and effects before he begins his work and combines substances."[8] The tradition of natural history, mentioned earlier – namely, the story of nature in terms of the kinds of natural objects and their order in identification – is also based on naturalism: we grasp "what is" by viewing its outer appearance from the perspective position in which the person as observer forms the centre.

Positivism is the second fundamental element of science. According to Auguste Comte, all knowledge throughout human history passes through three stages: the theological, the metaphysical, and the positive. In the theological stage, events are accounted for in terms of the powers of God or gods. The metaphysical stage deals with what lies beyond the physical, with essence beyond experience. It concerns itself originally with the knowledge of ultimate reality, which is derived from reason alone. Finally, the positive stage limits itself to bodily facts and their lawfulness, testable by means of sensory experience. Once the positive stage has been attained, according to Comte, the other views become antiquated and are eliminated. Today, the people of Western civilization find themselves in the positive stage.[9]

In addition to naturalism and positivism, objectivism is a necessary ingredient of science. Objectivism has as its aim the construction in thought of an outer reality that is independent of the person. According to Albert Einstein, observers have a strong effect on what they study, and thus it is necessary to eliminate the effect of the observer. The principal factor contributing to this subjective element is the evidence provided by the senses.[10] Hence, a theory is a system structured by reason, which approximates a true description of one objective external world. Following the experimental measurement, the observed system is left alone again, separated from the observer. So the world of sensations must, according to Einstein, remain merely an apparent world in contrast to the real world.

Progressive emancipation of thought from the data of immediate experience is most conspicuous in the replacement of the thing-concept by the concept of law. We obviously need, therefore, not only to develop an inventory of nature but also to know and apply the laws of nature. Scientific conception has as its aim the discernment of a common objective lawfulness underlying external reality. Hence, the thing-concept does not suffice. We do not merely wish to know "what is"; we also wish to know how items are linked together lawfully.

The lawful relationships cannot be observed as such; they require for their elucidation a particular kind of conceptual structure, which permits their construction in thought. Structure within science is based on induction and deduction. "Induction" means that concepts, statements, and judgments arrived at through analysis and synthesis apply not only to particular individual cases but to all objects and cases of the same or a similar kind. This assumes the uniformity and lawfulness of nature. Induction is coupled with deduction; "deduction" means the inference from the general to the specific case. From the general, we deduce particular consequences that can be tested by means of observations.

In science in general one is dealing with relations. More particularly, one is dealing with the relations between sets, where a set is a collection of objects or elements: $x1$ is an element of the whole set of $x$'s, $A$. In the relationship between the two sets, $A$ and $B$, given a certain value of $x$, one may unequivocally identify an associated value of $y$, which is the variable element of the other set, $B$. This association is known as a "function."[11]

In order to make predictions about our experience in the future of natural events, we must create inner symbols of outward objects. The relationship of the symbols must reflect the relations between the objective things and the way they depend on each other. Chemical equations illustrate this. The equation $CO_2 + H_2O = H_2CO_3$ represents in symbolic form the process of photosynthesis as it takes place in green plants. In this process, the carbon dioxide ($CO_2$) of the air is combined with water ($H_2O$), and the two compounds are transformed into the carbon-containing organic materials ($H_2CO_3$) of the plants carrying out this process. The symbolic structure in general must be such that the inference from the image or idea is an image of what naturally and in sequence follows the object that is its basis. When we explore the factual order of the sequential events and formulate them sequentially, we obtain the scientific structure of knowledge.

To illustrate the functional relationship between two sets by means of the example just given, one set may be the carbon dioxide of the air surrounding a green plant, variable in terms of its concentration; the other set may be represented in the form of photosynthesis performed by the green plants, measured as carbon-containing organic materials ($H_2CO_3$), which the green plants produce in varying rates. The rate of production of carbon-containing organic materials ($H_2CO_3$) may then be a function of the carbon dioxide ($CO_2$) concentration of the air.

General expression of functional relations is not possible by means of ordinary language; it requires the symbolic structure of mathematics based on number. Changes in number are expressible as a function of a homogeneous time scale. Accordingly, in mathematical natural science, one reaches the highest level of determinateness in terms of the laws underlying the material phenomena of nature. It must, therefore, according to Ernst Cassirer, be the aim to set thought free from immediate experience and to find instead a structure of thought that is based on the invariants of experience.[12] Because such a structure of thought is concerned with functional relations, number has to take the place of vision. While vision, or its representation in the form of the picture, portrays particular spatial compositions, reflecting immediate experience, our thought must concern itself instead with conceptual mathematical space.

Any science, then, according to F.S.C. Northrop, first passes in its normal development through a natural history stage, to which belongs a concept of inspection. Once this stage has been fulfilled, a science enters its second stage. This is referred to as the "postulationally prescribed theory," to which belongs a concept of postulation.[13] The meaning of such a concept is prescribed for it by postulates of the deductive theory in which it is embedded.

The modern study of the organic world wishes also to become theoretical. The science of ecology acknowledges only two principles in nature: energy flow, and the circulation of materials. Energy flow takes place through the food chain, which is a sequential process of eating and being eaten. Accordingly, one distinguishes two biotic components constituting a unit of nature: an autotrophic component capable of making food from simple inorganic substances taken up from the environment, and various levels of heterotrophic components, which cannot make food as autotrophs do but which must utilize materials already manufactured by autotrophs. For the ecologist, the organic world is thus essentially composed of only two constituents, producers and consumers. The ecologist speaks about animal production and plants as producers, while you and I become merely consumers. The principal steps of producing and eating, and of being eaten, are referred to as trophic levels. There are various kinds of trophic levels, such as the autotrophs, or primary producers, and the heterotrophs. Among the heterotrophs are those that eat plants – the herbivores – and those that eat other animals – the carnivores. The herbivores and carnivores are also referred to as primary and secondary consumers because of their respective positions within the food chain. But there are further trophic levels of consumers, and we thereby arrive at a functional classification based on trophic levels. This represents an abstraction from the visible into the realm of numbers, because production is a rate, expressed, for example, in terms of calories per unit of space and time.

### SUMMARY

Up to this point, I have discussed the current objective view of the natural world. This view holds that reality is composed of two items only: the observer and the observed, which are also, respectively, the knower and the known, or the subject and the object. The role of the subject is that of the impartial watcher, the thinker, whose task is to gain knowledge of the world. This knowledge is to be objective: it must be true, regardless of the person acquiring it, and it must be explicit, fully specifiable. In gathering this knowledge, the initial aim is to amass an inventory of nature in terms of the objects found in it. Studies of this kind are collectively referred to as natural history. One aspect of natural history is the representation of the objects of nature – for example, through pictures. This in turn calls for the central perspective projection in which both the observer's position and the position of the observed are fixed and the element of time is eliminated.

The natural history stage of science is followed by the stage of the "postulationally prescribed theory" of nature. One does not merely wish to know "what is"; one also wishes to know how the objects are linked together lawfully. This theoretical stage of science allows one to discover the laws underlying the phenomena of nature. While the picture representing the object of nature is central in the natural history stage, in the theoretical stage the picture is replaced by the number.

### CHAPTER TWO

# Human Beings and Nature in the Mythological World

## HUMAN BEINGS IN THEIR INDIGENOUS STATE

In forms of mythic apprehensions, we learn how something came to be, its origins and emergence, and how various aspects of the world form a whole life connection, including humankind. In the mythical experience, human beings and the world form a unity, and there is not yet an "I" separate from a "not-I." In this case, one experiences space from within rather than from without. Space and space content belong together. The part is a function of the whole, because the part does not exist by itself but is linked to a centre from which it receives its structure and its significance.

In order to understand the nature of myths, we need to distinguish with the sociologist Mircea Eliade between two very different experiences of space, the experience of profane space and the experience of sacred space. Eliade defines the sacred as the opposite of the profane. For the manifestation of the sacred, he uses the term "hierophany," meaning "that something sacred shows itself to us."[1] The experience of sacred space is made possible through the *sensus numinis*,[2] which is the resonance of the soul to the supernatural in which the soul participates. For indigenous people there is an absolute reality that transcends this world. But that reality reveals itself in this world. The sacred may manifest itself, for example, in a tree. Eliade emphasizes that in this case the sacred tree is not worshipped because it is a tree but because the tree is a hierophany, showing something that is sacred – a supernatural reality. Accordingly, any aspect of nature can reveal itself as sacred.

In an indigenous society, the relationship between the member of that society and the world is very different from our own. The anthropologist-philosopher Levy-Bruhl has found that many indigenous people depict the world in the form of a collective representation.[3] According to him, "Collective representations are social phenomena, like the institutions for which they account, and if there is any point which contemporary sociology has thoroughly established, it is that social phenomena have their own laws, and laws which the analysis of the individual qua individual could never reveal. Consequently, any attempt to 'explain' collective representations solely by the functioning of mental operations observed in the individual (the association of ideas, the naive application of the theory of causality and so on), is foredoomed to failure."[4]

Levy-Bruhl warns us, though, not to understand by collective representation one that is merely cognitive, for it is interwoven with emotional and motor elements. He makes a clear distinction between the collective representations of indigenous people and our own predominantly cognitive representations. He refers to the former as mystic representation, by which he means "belief in forces and influences and actions which, though imperceptible to senses, are nevertheless real. In other words, the reality surrounding the primitives is itself mystical. Not a single being or object or natural phenomenon in their collective representations is what it appears to our mind."[5]

For native people, it is the communication with the invisible, intangible, and mutable forces that is important. In the collective representations of native people, particular phenomena may give rise to certain forces, such as demonic forces. These forces, arising coincident with the phenomena, remain a part of the phenomena; they are not separate. Unchangeableness of things is not important, therefore, since each knower interprets the forces differently. The phenomena in their collective representation are not lawful in themselves but depend on changeable relations between the group and aspects of its surrounding. The mystical forces thus exhibit continuity from thing to thing and are related to each individual's own nature. There is, for example, an identification between the individual members of a totemic group and their totem. Levy-Bruhl terms this the "law of participation." Everything that happens influences at the same time both the individual as a member of the group and the surroundings of that individual. The law of participation, or "participation mystique," implies that no clear distinction is made between the "I" and the "not-I."

The religious person experiences absolute reality beyond this world which manifests itself within this world in the form of sacred place. For the religious person, space is not homogeneous, because holy ground may be surrounded by formless expanse. Indigenous society needs to create a place out of the experience of sacred space. The image of the perfect place is that of paradise. Paradise is the sacred garden – it is created by God: "Then the Lord God planted a garden away to the east, and there he put the man whom he had formed. The Lord God made trees spring from the ground, all trees pleasant to look at and good for food; and in the middle of the garden he set the tree of the knowledge of good and evil."[6]

A garden is an extended piece of ground, cultivated with plants as the elements. These elements are placed on the ground in the form of a composition, thus creating a unit of a higher order. The plants themselves represent life-forms that cannot be understood merely by themselves but must also be viewed in their relationship to other life-forms. These relations we call "correspondences." Take, for example, the life-form of circumboreal trees, such as birches, and that of rhizome plants, such as the *Anemone nemorosa*, which overwinter hidden underground, forming leaves and flowers above ground early in the spring before the birches unfold their leaves and cast shade on the ground. Within the composition, the elements are fitted together according to their correspondences, one in relation to the other. In the plan of the Garden of Eden, all these correspondences are taken into account so that the greatest possible sense of meaning emerges among the life spaces within the garden as a whole. This includes the gardener himself, Adam, whom God asks to take care of that most harmonious of all gardens.

The garden and the gardener in their full correspondences owe their existence to God. In this context it is evident that God is most deeply concerned about humankind. His concern for the gardener expresses also his concern for the garden, because he has created the gardener and charged him with maintaining the perfect state of the garden as God created it. For this reason, God also brings the tree of the knowledge of good and evil to the special attention of Adam both by pointing it out to him and by forbidding him to eat of it: "The Lord God took the man and put him in the garden to till it and care for it. He told the man, 'You may eat from every tree in the garden, but not from the tree of knowledge of good and evil; for on the day that you eat from it, you will certainly die.'"[7]

## SEVERANCE FROM THE GARDEN: THE STATE OF WILDERNESS

In Genesis, we are told that it is the snake that makes Adam and Eve eat of the forbidden tree and that the snake is by far the most crafty of any of the wild creatures found in the garden. As soon as Adam and Eve eat of this tree, their self-awareness and self-reflection are born; they lose their unconscious ability to enter the "outside" by participating in it, as they had formerly been able to do within the context of the experience of sacred space. Human beings, as we know them have been born, having gained a degree of self-awareness: "Then the eyes of both of them were opened and they discovered that they were naked."[8] From that moment, humankind finds itself in a vast movement away from the state of innocence and towards a growing self-awareness. This movement requires that one discover oneself as an individual, rather than collectively, and that one now discover the "not-I" as separate and apart from the "I."

Before the Fall, Adam's skill as gardener depended on his collective unconscious ability to enter into that "outside," the garden, by the law of participation. Through the law of participation, members of indigenous society identify their own life with the life of aspects of nature. To achieve the distinction between the conscious "I" and the "not-I" in the form of the unconscious nature of plant and animal life, it is necessary to challenge this state of being linked to nature. The totemic connection must be severed. In our Western religious heritage, the emancipation of the human soul from nature is crucial. Earth becomes darkness and the personal god is now the god of light. The gods of the earth, as the expression of the unconscious, are banned in order to establish more firmly the ethical principles founded on rational thought.[9] According to the Western religious tradition, God created humankind in his image. Through the Fall, however, we have lost this image. According to the Christian Church, God has renewed this image within us through Christ. The redemption takes place by reuniting us with the image of God, and it is thereby a, restoration of the original state. We need to be reminded that images as forms of relations may guide us in our interpretation of the world.

In antiquity, the soul was viewed as a natural object and therefore as part of nature, part of the cosmos. Consequently, being alive and possessing a soul were identical states, because the soul gave the body its form and its movement; it was the invisible vital force. With the event of the Latin church fathers, the outlook on the world changed profoundly. Whereas the soul had previously been considered part of nature and not therefore separated from matter, Christian doctrine holds that the soul is superior and beyond nature. Thus, the soul now lifts itself out of the structure of nature because it has been created in the image of God. The soul is the gate to the religious life, since it is through the soul that the union of the believer and the believer's personal god is realized. But the soul has to become completely removed from all the connections with nature that existed in antiquity, and this makes it necessary to renounce all the things of the world because they interfere with the awakening of the soul.[10] Augustine wondered how it had ever been possible that people had recognized only the external world, as in antiquity – that they had admired the cosmos rather than looking inside themselves; for we are able to experience our soul directly, and what is certain, therefore, is not what is outside but our own soul inside. Through our attention to our own soul in relation to God, we are able to discover our very own inward dimension.

An example of the conflict between giving one's attention to the external world and gaining one's inward dimension is borne out in an experience of the great humanist Petrarch. In a letter of 26 April 1336, Petrarch told Dionigo da Borgo san Sepolchro, professor of theology in Paris, about his troubles while climbing Mount Ventoux.[11] As soon as his soul opened itself up to nature in the profound experience of the magnificence of the landscape as he viewed the prospect from the summit of the mountain, he sensed that he had fallen victim to the earth. It had tempted him. He opened Augustine's *Confessions* and hit by chance on the following sentence: "And men go to admire the high mountains, the vast floods of the sea, the huge streams of the rivers, the circumference of the ocean, and the revolutions of the stars and desert themselves."[12] This sentence reminds Petrarch that his inward dimension is very fragile in its nature and that he may be abandoning it altogether at the moment he begins to admire the external world, such as the view from the mountain. In the face of what he senses to be a temptation, he submits to the ideas born within his own inward dimension and again abandons the external world. We can see here how the discovery of the inward dimension took place at the cost of focusing less and less attention on the external world. The individual's life thereby became isolated from things and beings of nature as they are within their own settings and as they might otherwise have begun to form integral parts of the individual person's circumstance.

So antiquity does not recognize our potential inner dimension; and Christianity, while creating that dimension, has also prevented us from recognizing the world in its own being. Since the external world potentially interferes with our inner life, it merely serves to satisfy our everyday bodily material needs. It is relegated to materiality; the utility of bodily things for our material life may be taken as a criterion for truth regarding nature and hence as a measure for the order of things composing nature.

Now, the expulsion from the Garden of Eden has had a profound impact on the characteristics of human beings. Simultaneously, it has impaired the state of the garden. When Adam and Eve were driven out of the garden, there was no longer anyone present to take care of it. At that very moment, the state of nature was born. Wilderness, spreading from within the garden, is nature in the absence of humankind.[13] We have all experienced the state of wilderness and we retain the image of wilderness: the tract of land without paths, the darkness of the dense forest, the roar of the wild animal. The image of the wilderness is expressed, for instance, in Henri Rousseau's description of his painting *Exotic Landscape*: "The hungry lion tears at the antelope, the female panther waits for a share of the booty, the bird in the tree is already holding a piece of meat it has snatched in its beak, the suffering animal sheds a tear. Then the sun goes down."[14] The spaces of wilderness are dominated by wild animals. By their very nature, the animals care only for themselves and their kind. From this condition arises an innate struggle; enemies and danger become inevitable aspects of each life-space.

Because wilderness is the place of the wild animals, it is not our world. In the presence of the wilderness we get a sense of our own powerlessness. Nature in the state of wilderness is perceived as dangerous space and produces in us a deep sense of fear. All of us have experienced this fear, fear of the deep forest at night, fear of becoming lost in such a lifespace. Wilderness is God's curse on the human race: "I will fling you into the wilderness, you and all the fish in your streams; you will fall on the bare ground with none to pick you up and bury you; I will make you food for beasts and for birds."[15] Wilderness is an obstacle to self-realization, because nature in the state of wilderness gives rise to evil spirits and to seduction by one's bodily desires. Human beings in their state of savageness worship magic forces in the darkness of the wilderness, and it is these forces that come to be viewed as the forces of evil.

Because wilderness is the place of the wild animals and not our space, there exists no space for us. The philosopher Otto Friedrich Bollnow points out that the English "room" and the German *Raum* may be used in the same manner, as in the expression "to make room," which means that room or space is not present as such.[16] Room is gained only through our activity. Wilderness is not space as such, but room is gained from the wilderness through clearing. Room refers to our own life; room exists only in our own world, and since wilderness is not our world, it does not possess room.

### SUMMARY

This chapter looked at the relationship of humans and nature in the world of indigenous people and found that for the religious person there is an absolute reality that transcends this world. This reality reveals itself in the world in the form of the sacred. In totemic systems of indigenous people, for instance, the distinctive qualities of the totem animal manifest its sacred nature. Each generation is the renewed embodiment of the animal ancestors. In the ritual ceremonies of indigenous society, these ancestors are actually present. There is therefore a unity of the present generation and the totemic animal, and they are thus both members of the same world. Indigenous people are directly linked to nature in this manner, and no clear distinction is made between the "I" and the "not-I."

In order to achieve such a distinction between the conscious "I" and the "not-I," the totemic connection has to be severed. The reverence of the ancestors has to be replaced by the sense of awe in the presence of the holy god. God becomes the concrete, living, and personal god of the Old Testament. God is the god of light, and the unconscious animal-like part of our existence is viewed as darkness which needs to be superseded by the conscious life, because only that life is illuminated by God's word. To achieve this status, nature has to be separated from us – we cannot remain within it. The gate to the divine life is the soul. In antiquity, the soul was an object of nature and therefore part of nature, part of the cosmos. In the Christian faith, the soul must raise itself above nature and become one with God. The soul is a person's inward dimension. In order to nourish the soul, one has to focus one's attention fully on it in its relation to God. Only then is one able to attain awareness of the "I" and at the same time attain distance from the "not-I." Because the things of the world interfere with the awakening of the soul, nature is demoted to a mere object and a resource for our material life.

### CHAPTER THREE

# The Interpretation of Humankind

## HUMAN AFFECTATIONS

The expulsion of Adam and Eve from the Garden of Eden has had a profound impact on human nature. The importance of the Fall is described for us in Kleist's "Marionette-Theatre."[1] In this short story, Kleist lets us listen to a conversation between himself and the first dancer of the opera house, who loves to watch the marionette theatre in the marketplace. The dancer compares the grace of the puppet's dance with that of a human dancer and comes to the surprising conclusion that the dance of the jointed doll is much more graceful than that of any human. The human dancer cannot avoid blunders, since humankind was born when Adam ate of the tree of the knowledge of good and evil. Through this act, the human race acquired affectation, which represents a state of being in which the position of the soul does not coincide with the centre of gravity of the person's motion.

By contrast, within each animal the position of the soul and the position of its centre of gravity coincide; as the animal moves about and acts in relation to meaningful signs in its surrounding world, the centre of gravity follows the same path that the soul describes. The first dancer in Kleist's story describes a fencing match between himself and a bear. In this match, the bear turns out to be the master because the seat of the soul and the centre of gravity coincide within the bear, whereas this is not the case with the dancer. Through eating from the tree of knowledge in the Garden of Eden, humankind has become self-aware, in contrast to animals.

The dancer also tells of a young man who destroys all his former grace as a result of inevitable forces brought into motion through affectation. As soon as the young man becomes self-aware, the locations of his soul and his centre of gravity move apart because it is human nature to attempt to assume or exhibit what is not natural. Accordingly, Kleist concludes: "Thus at one and the same time [gracefulness] appears in its purest form in that human body which has either no self-awareness at all or an infinite self-awareness – that is, in the jointed doll or in the god. 'Therefore,' I said a little absent-mindedly, 'we should have to eat again of the tree of knowledge in order to fall back into the state of innocence.' 'Indeed,' he answered, 'that represents the last chapter in the history of the world.'"[2]

Kleist illustrates our afflictions as they result from the Fall. Before Adam eats from the tree of the knowledge of good and evil, his sense of existence in sacred space expresses itself in a profound relationship between the sacred garden and the spirit of the gardener; the garden forms an extension of the gardener himself. This mode of participation requires the openness of human innocence towards people and among other creatures. As soon as this openness within the collective breaks down and secretiveness arises in the individual, the knower of the secret is estranged from the collective and a sense of guilt develops. This danger to the innocence of humankind is symbolized by the tree of the knowledge of good and evil.

### THE HUMAN BEING AS THE RATIONAL ANIMAL

In the modern world, people have lost the sense of sacredness because they have gained the sense that they are free to make themselves and their world. The tool enabling them to make their own world is the power of their thinking. Descartes viewed the human being as the rational animal, characterized by thinking. Descartes taught us to doubt that the senses can give us access to the "being" of nature. He considered modes of knowledge based on experience to be inferior to modes of knowledge based on thinking. According to Descartes, there is something that wants to deceive us (indeed, there is an evil demon that makes us err), and our mind is especially deceived by our senses. We must therefore begin by doubting everything and in particular the information obtained by the senses. Descartes distinguished between the world of experience, based on our own senses, and the world of science, which we construct by means of thinking: "I would have you note the difference there is between the sciences ... and in general all that rests on experience."[3] The structure of "being" arises within consciousness itself in the form of thinking.

The basis of thinking is reason, and more particularly it is mathematical reason. We have very limited access through our senses to the lawful relations that pervade the natural world; these relations have to be constructed in thought. The items of the external world and their perceived relations must be transformed into patterns of mental symbols, such as mathematical ones. Such ideas as extension and motion of bodies give us access to the laws underlying the phenomena of nature. Mathematical reason links together in a lawful relationship the sequential positions of bodies in space. This form of thinking, then, allows us to calculate the mechanics of the world.

Descartes believed that the whole of reality is made up of two substances, both created by God: thinking substance and extended substance. The essence of the material realm is extension, while the mind is the realm characterized by the act of thinking. Reality is thus ultimately twofold in the form of *res extensa* and *res cogitans*. That which is extended is not conscious and that which is conscious is not extended. Descartes established thereby a complete dualism of mind and matter. In the Cartesian view, there is the world of material bodily externality and the world of consciousness. The mind needs to be separated from the body because mind and body belong to two different worlds. The mind is composed of thinking substance. Thinking is separate from the bodily things. Knowledge of the world is thus twofold, composed of explanatory knowledge of the external world based on physical reason and on the ideas arising within human consciousness.

## THE NATURALISTIC SCIENTIFIC VIEW OF HUMAN BEINGS

In the first chapter I described the objective view of the natural world. Now I want to ask whether one can also view humans within that same context. In discussing different orders in the natural world, I contrasted a systematic order with a spatial order. A systematic order is to be based on natural relationships that can be elucidated by means of comparative investigations into the fundamental structural plans of different organisms. Structural sameness within the architecture of different organisms is called "homology"; homology is not concerned with similarity or dissimilarity of function.

Different organisms that correspond to each other in their fundamental structural plan may exhibit very different forms. For example, if we compare the skeleton of a frog with a human skeleton, we discover complete similarity in the basic plans of these structures, in spite of great differences in the size and relative proportions of their component parts. We can therefore state that the human skeleton is homologous to the skeleton of the frog. The great differences between frog and human in their respective forms are modifications of one and the same structural plan. The spatial order, such as that discovered by Humboldt for plants, may arise through different kinds of organisms attaining similar life-forms, even though their fundamental structural plans may be entirely dissimilar. By "life-form" is meant the design of the organism in relation to the performance of that organism's basic ecological functions, for instance, the function of locomotion. Similarity between different organisms in terms of their life-forms is known as "analogy." Analogy is of great importance in grasping order in the diversity of life-forms in relation to their spatial features. It is evident, therefore, that two organisms or their parts may be remarkably similar in their appearance but dissimilar in their underlying fundamental structural plans. The wing of an insect is "analogous" to the wing of a bird. Two organisms and their component parts may, however, be dissimilar in their appearance and function, and yet they may be strikingly similar in their underlying plans. A man's arm and a dog's foreleg are "homologous" but not analogous.

Viewed from this standpoint it becomes apparent that human beings, too, fit into this natural systematic order. On comparative grounds, based on the external bodily appearance and especially the internal structural organization of humans and certain kinds of animal, within the diversity of animals a natural generic order can be established in which humankind has its place too. Accordingly, the human is an animal belonging to the phylum Chordata, to the class Mammalia, and to the order of the Primates, where on the basis of bodily structural similarities humans are placed together with the monkeys, lemurs, chimpanzees, apes, and gorillas. Within the order of the Primates, humans are considered to belong to the separate family Hominidae and to be the only living species belonging to the single genus *Homo sapiens*. In this argument, we have adopted the naturalistic scientific position, the view that what is most real about any phenomenon is its physical aspect. The external world is composed of bodily matter existing in three-dimensional space. Matter is the only element of the extended world; nature is viewed as existing only in this form.

The zoologist Hans-Wilhelm Koepke has pointed out that it is the life-form, the features of the organism, that determine how the organism carries out its basic ecological functions.[4] The manner in which the organism performs its basic ecological functions in turn determines the spatial order in diversity. Each life-form belongs to a particular life-space, the "habitat." The nature of the habitat is to a large extent determined by the features of the life-form. Wherever there is a high density of specimen habitats of the same life-form, we are dealing with a "habitat maximum."[5] In a bog, for example, the most abundant life-forms are peat moss, bog-pink orchids, and round-leaved sundew, indicating particular habitat maxima. These forms may be absent in neighbouring drier areas where there are other habitat maxima instead. An area where a number of habitat maxima come together in a coherent manner is known as a "biotope."[6] By biotope we mean the space as it is characterized by the various prevailing abiotic conditions and by the most frequent habitat maxima. Characteristic life-forms are present in one kind of space while others are absent, though they may prevail in a neighbouring area.

Koepke has designed a broadly arranged classification of biotopes, consisting of: (1) space biotopes, which are the interior of a medium; (2) plane biotopes, situated at the interface between two media; and (3) shore biotopes, those at the interface between three media.[7] In a savanna, an area characterized by a high density of specimens of grass life-forms, acacia trees, et cetera, the grazing animal in its setting lives in a plane biotope – the interface between two media, which in this case are a solid medium and a gaseous medium (the solid ground on which the animal moves and the air surrounding it).

Humans are like the animals that live at ground level. We move about, for example, through the openness of the savanna. In general, live human bodies are much of the time bodies in motion, as are those of many different kinds of animal. The human body has the structural plan of any vertebrate body, distinguished by the presence of a vertebrate skeleton. The skeleton serves both in supporting the body and in the movements of its parts.

Organisms are bodies, physiological bodies that end with the outer boundary of the body. The body in turn is seen as a machine; and as a machine is composed of parts, so is the organism's body, and these parts are integrated into the whole entity. This integration is accomplished through specific relationships of the parts to each other. As the machine is defined by its operational principles, so also is the organism defined. As the machine is designed to perform certain tasks, so is the human body. At its lowest level of performance, the body carries out such basic functions as sleeping, respiring, and heat regulating. The body's execution of these functions is necessary so that it can maintain its capacities at the vegetative level of human life, upon which are built other levels.

How, though, should we conceive of the relationship between any organism, including the human being, and its environment? Such a relationship is thought of in terms of stimulus and response. Everything acting from within the body, demarcated by its outer boundary, is conceived of as being intrinsic, and that acting from without is conceived of as extrinsic. The interaction between intrinsic and extrinsic forces is thought of in terms of irritability. Accordingly, the organism is irritable matter; irritability is therefore intrinsic. The extrinsic forces are defined in terms of stimuli, which are physical forces, such as light. The external visible world of perception is therefore identical with the world of physical light phenomena as they are studied by the physicist. The student of psychophysics Gustav Theodor Fechner inquired, for instance, into the functional relationship between the effect of the stimulus on the reactive system and the intensity of the stimulus. He was interested in constructing a scale for measuring the perception of brightness that could relate mathematically to the physical scale of light intensity.[8] Hence, the relationship between extrinsic and intrinsic forces follows a natural law that can be stated in terms of measurable properties.

How is the body able to reflect environmental reality? The anatomist thinks in terms of the "principle of division of labour." This principle states that particular organs perform specialized functions and that generally there exists a one-to-one relationship between an organ type and its function. The anatomist distinguishes sense organs from digestive organs, and among the sense organs a distinction is made between those sensitive to touch, those responsive to vision, and so on. The organs are structurally specialized to carry out specific functions, acquiring particular kinds of information from the external world. They receive this information in the form of stimuli, which bring about excitation of their component sensory cells. Touch cells are excited by pressure stimuli; the sensory cells of the eye are stimulated by light.

The anatomist also separates the sense organs from the nervous system and identifies the principal task of the nervous system as the conduction and storage of the excitations produced by the external stimuli. For example, the stimulus of light is received by the eye but is transmitted by the nervous system. In more highly evolved animals, the brain forms the central part of the nervous system at its anterior end. Information about the external world is obtained by the eye and transmitted to higher centres of the brain as a set of corresponding points. By means of the eye and the nervous system, the physical realm of the external world is transformed within the brain into its reflection, similar to the reflection of the exterior scene captured by the film in a camera.

The Russian physiologist Ivan Petrovich Pavlov studied the direct response of an organism to the environment in terms of a dog's reflex by which, in the presence of food, the mouth and the stomach secrete their various digestive fluids. The presence of food in the mouth represents the stimulus, and the secretion of the fluids is the response. We can view this also as an aspect of a system that potentially contains an input, an output, and a set of rules governing the relationship between input and output. The input in this case is, of course, the presence of the food, while the output is the secretion of the various fluids. There are systems in which the output has no influence on the input, but there are also many systems in which the output does influence the input. The latter will generally contain feedback. By "feedback," we mean that the output influences the input of the system, which in turn has an effect on the output.

The body has stability. It is capable of maintaining internal conditions within a certain limited range when external conditions fluctuate. Feedback control enables organisms to maintain these fairly constant internal conditions. In 1948 the American mathematician Norbert Wiener published his *Cybernetics – or Control and Communication in the Animal and the Machine*. Wiener called the entire field of control and communication theory (whether in the machine or in the animal) "cybernetics," which means steersman.[9] The concept of information is of great importance in this context; and closely associated with the concept of information is the concept of communication.

When two people communicate, something is transferred from one to the other. By communication, we therefore mean the transfer and exchange of information. In communication, information is transmitted in the form of signals or signs from a bearer to a receiver. Communication is not restricted to humans; it is equally applicable to animals. In the case of humans, communication may be in the form of language, which may then be transduced into a signal, a code (for example, the way the letters of written language have been transduced into the Morse code); and it must be transferred through a medium to the receiver, where it is initially present in the transduced form before it is decoded. Similarly, environmental stimuli are coded in the receptors of the body into electrochemical signals of the nervous system and are conducted to the brain, where they are decoded. The information can then be stored, but stored information may not remain unchanged.

It often happens that new information becomes linked with stored information in the form of associations. In Pavlov's investigations of his dog's reflexes, he showed that it was not only the presence of food in the dog's mouth that induced the secretion of fluids. The dog also started to secrete when he merely saw the food. Pavlov then showed that the dog could be made to salivate in the presence of other external stimuli, such as the sound of a bell. Initially, the bell was rung when the food was given. After a time, this new stimulus could bring about the secretion response in the absence of the food. Such a connection between a new stimulus and an existing reflex process is known as "conditioning." The sound of the bell is paired with the food. The dog learns the connection between the sound of the bell and the food. Such physiological investigations dealing with reflexes – with direct responses of the organism to the environment – greatly promoted the mechanical interpretation of the relationship between the body and the environment. In this case, even the phenomenon of the psyche can be accounted for in terms of a physiological information-processing mechanism.

### SUMMARY

This chapter examined the nature of human beings. The Fall has had a lasting impact on it. According to Kleist, as a result of the Fall, humans acquired affectation, and were thus no longer capable of acting naturally. They became self-aware and attained a critical attitude. It was this attitude that made Descartes doubt everything. The only item he was unable to doubt was his own thinking. The ability of humans to think is made possible through the presence of the mind. The Cartesian view holds that the human being is the rational animal and that thinking is separate from bodily things. The world is thus composed of two different principles: mind and body. Mind is thinking substance and body is extended substance. Mind is a substance without any extension and is characterized by the trait of thinking. The extended world of bodies is explained in mechanical terms. Descartes identifies the law of nature with the principles of mechanics. While this principle applies equally to animals and human bodily functions, only human beings have a thinking substance in addition to extended substance.

Science acknowledges a world that is composed solely of bodily things. In order to investigate humans scientifically, the human body must be compared with the bodies of various kinds of animals. There are striking similarities between human bodily structure and the bodily structure of some kinds of animals. On these grounds, we may conclude that humans are simply another kind of animal. In addition, when one seeks to account for organism-to-environment relations, one finds the stimulus-response model to be equally applicable to both animals and humans.

### CHAPTER FOUR

# The Historical Nature of Humankind and the Contemporary Technological Order

## THE HUMAN AS A HISTORICAL BEING

Animals are part of nature, and since humans are just another animal, they are in nature and not apart from it, so the argument seems to unfold. But according to Descartes, the human being is the rational animal. In addition to the extended world, there are the ideas arising within human consciousness. From the historian's perspective, the horizon that lies open to human consciousness is not fixed; it changes with history. Thus, the human is a historical being, and it appears that there is not human nature as such; there is solely historical mentality. In addition to the physical world to which we have access through physical reason, there is a reality, humans themselves, a reality that can be revealed in the form of historical reason. Consequently, we today recognize the humanities as forming a system of knowledge that is based on historical reason and is independent of the natural sciences. In the natural sciences, one explains in terms of physical reason, but in the humanities the aim is to understand a person's motivations within the historical context.

For those of us who acknowledge explanation as a source of our thinking, understanding is a new term in our vocabulary. Furthermore, the question arises: Understanding of what? According to the psychologist Eduard Spranger, we as human beings are capable of understanding only other persons.[1] The particular meaning a person attaches to things and beings is determined by the kinds of forces arising within that person – for example, economic, biological, and aesthetic forces. In order to understand, historians must acquaint themselves not only with the basic forces originating within a person but also with the specific situation in which a person acts. For this, historians need to gain as complete a picture as possible of a particular historical constellation and the character of the acting person. By "character" is meant the scale of values which that person uses in appraising the composition. Through imagination, historians must comprehend the corresponding divergence between their own experience and the other person's, what Spranger refers to as *mutatis mutandis*: necessary changes having been made.[2] This includes the image of the surrounding world of the person whom a historian wishes to understand, and it relates the composition of the surrounding world, in its very concrete form, to that person.

Individual acts are guided by values that the individual projects into the composition. The historian, in trying to understand a historical figure within that person's cultural world, must judge that figure neither solely in terms of the acting subject nor in terms of the object or situation to which the subject is relating. The historian must take a position somewhere between the two. Understanding encompasses both subject and object, and the search for the meaning of an act requires that the principal projections of the subject's mind as well as their underlying structure be elucidated.

As human beings, we potentially have access to these fields of projection because we all participate and reflect on the same world in terms of our own being. We are able to understand the acts of other persons by dwelling in the acting person. We can gain a concrete picture of the acting person's situation through our own experience, appreciating the basis for the person's act, as if it had arisen within ourselves. This is possible because our lives have a common foundation.

I stated above that the horizon that lies open to human consciousness changes with history. A particular cultural world has its very own characteristic motives. In the context of our own culture, for instance, and with special reference to the Middle Ages, the historian-philosopher Wilhelm Dilthey has identified three principal motives that were combined in particular ways during this epoch. These are the religious motive, the aesthetic-scientific attitude of humans, and the task of human will.[3] The core of the religious motive is the relationship between the human soul and the living God, which must be experienced in order to be understood. In their aesthetic-scientific attitude, humans concern themselves conceptually with the logical connections of the world as a whole, how it hangs together, expressed in the form of explanatory knowledge, which becomes the criterion for the interpretation of perception. Finally, the task of the will is to shape and regulate worldly empires and their relations to the life of the individual.

Since the beginning of the Renaissance, these three motives have no longer been interwoven, for the spirit of the Renaissance was dominated by the aesthetic-scientific attitude. The need to shape the environment is now foremost so that it becomes separate from and subservient to human life. The only motive relevant to this aim is the scientific one. Accordingly, it begins to dominate the individual. The consequences of the elimination of the religious motive are illustrated to us by Friedrich Nietzsche. In *The Joyful Wisdom*, Nietzsche describes a situation in which a madman lit a lantern in the bright morning hours and cried incessantly in the market place, "I seek God! I seek God!" Many of those who stood around did not believe in God and accordingly the madman caused a great deal of laughter. The madman then called out, "I mean to tell you! We have killed him – you and I! We are all his murderers!"[4]

The psychologist Julian Jaynes has described some of the consequences of the death of God, or of the silencing of God's voice.[5] The voice told us what to do. When the voice disappears, we become frightened because no one any longer tells us what to do. Without the voice, we must learn to reason within our "mind-space" how to act in relation to a particular novel situation. Lacking God, we must constitute ourselves, because each of us is founded solely within ourself. According to Nietzsche, "In fact, we philosophers and 'free spirits' feel ourselves irradiated as by a new dawn by the report that the 'old God is dead'; our hearts overflow with gratitude, astonishment, presentiment and expectation. At last the horizon seems open once more, granting even that it is not bright; our ships can at last put out to sea in face of every danger; every hazard is again permitted to the discerner; the sea, *our* sea, again lies open before us; perhaps never before did such an 'open sea' exist."[6] The Book of Genesis tells us that God "created man in his own image." The concept of a specific human nature was the outcome of the idea of God the maker. The death of God brought to an end the meaning of human values held generally. We ourselves create all values and are solely responsible to ourselves for the choice of the values we make.

Like God's act of creation, we humans shape matter with our own hands. We may shape it into implements, for example. Our task is to shape the environment in the same way that we shape an implement. Technological human beings must mobilize the forces of nature for their own purposes in terms of technical production. The technological human, viewed from this perspective, is the consequence of the death of God.

From demonstrating on comparative grounds that the human being is an animal, to the interpretation of the nature of the relation between humans and their environment, the objective view of the world is all-pervasive. We feel the need to explain experiential phenomena in terms of correlated physiological processes or in terms of the objective structure of the environment. I stated earlier that we place ourselves in the centre, in the form of the central perspective projection. From this point of view, we discover objects based on their outer appearance, and we isolate items from the ground in which they are embedded in order to recognize what they are when seen by themselves. We think of them as beings by themselves, and we also think of them as independent of ourselves, because from our position at the centre we draw apart from the things in order to dominate them. Heidegger points out that we have seen the thing not as thing (what it is in itself) but as *res*.[7] He tells us that the Latin word *res* does not mark out what a thing is in itself; rather, it concerns how it applies to us in some sense. Furthermore, he thinks that we usually view our acts in terms of their utility. By "utility" we mean the ability to modify our environmental situation in such a way as to enhance what it affords us for our material life.

Nature contains powerful forces, such as the flow of the water of a river, the raging windstorm, and the rays of the sun. Since we wish to make these forces serve us in productive ways, we must learn to transform them into other kinds of forces, such as the flow of water through turbines, which we utilize to produce electricity. These forces – in this case, the flow of a mass of water and the mechanical work of the turbine and the electric power – must be revealed, and the nature of their equivalences must be brought into openness. This was the concern of the German physiologist Hermann von Helmholtz who, in a public lecture in 1854, described an iron hammer being driven by a waterfall.[8] Helmholtz moved quickly from the visual aspect of the landscape containing the waterfall to focus his attention on the force of the falling mass of water, and he showed how this driving force could lift a hammer and in turn lead to mechanical work by the hammer; the falling of the mass of water was translated into a measure of work. Helmholtz's interest was directed towards the equivalence of the inherent force of the waterfall and the mechanical work of the iron hammer. The amount of work performed by the falling hammer could be determined by the velocity of its fall. Thus, the motion of a mass represents energy that may carry out work. This energy is known as the kinetic energy of the mass.

Helmholtz emphasized that we can only obtain energy that is already present in nature; thus, nature represents an enormous storehouse of energy in the form of coal, water, wind, and so on. We can extract, steer, and exploit this energy for our own purposes, in the forms of mechanical force, heat, magnetism, or electricity. These transformations between the various forces require the invention of a corresponding conceptual framework, such as work and energy. We think in terms of human beings and animals performing work and machines doing mechanical work. In the mechanical sense, the concept of work becomes equivalent to the expenditure of energy. Helmholtz concluded his lecture by pointing out that relationships similar to those in machines also exist in humans. Since the body, for example, burns food in the same way as the steam engine burns coal, and since both produce heat, the body does not differ from the steam engine in terms of heat and energy. The body, too, has to be conceived as a machine that performs work.

In order to determine the equivalences, as Helmholz set out to do, we must view nature theoretically in terms of the postulationally prescribed theory. Functional relations are crucial. General expressions of functional relations are not possible by means of ordinary language but require instead the language of mathematics based on number. It becomes necessary to grasp phenomena of nature in accordance with causality. We call the causes of events in nature "force." We wish nature's forces to become productive forces. This requires a mechanical interpretation of nature in terms of the general definition of force as the capacity to do work. The forces arising within nature are to be channelled, as along the tracks of a railroad or in the wires transporting electrical energy.

The locomotive moves on its track by means of heat transformed into mechanical work, the direction of the movement being determined by the position of the tracks; it moves with a speed that is remarkably even compared with that of a vehicle drawn by animals. Steam power, as applied for example to a steamboat, is a force that causes the boat to prevail in its direct path largely independent of nature's forces, such as the forces of the winds that govern the spaces through which the boat moves. The resulting movement no longer depends on the nature of the space and the natural forces prevailing within it. Instead, the movement depends on the mechanical forces that create their own spatiality.

Out of nature in its original state we intend to create rational spaces. This requires that we alter the composition of our environment. Animals adapt to the conditions of their surrounding spaces, but humans adapt the conditions of their surrounding spaces to what they conceive to be their needs. By "need" we mean "a condition requiring supply or relief ... the lack of anything requisite, desired, or useful."⁹ We distinguish between subjective and objective needs. Based on certain images or notions, a need is subjective; based on essential connections, it is objective. To satisfy our needs with the least expenditure of energy, we have to produce objects (these being in contrast to natural objects). The produced objects assist us in our material life and make it possible for us to attain new lifestyles according to our will. Our ideas of well-being change – they may at one moment be the circumnavigation of the earth and at the next moment the eradication of disease. Whatever our idea of well-being may be, we want it to be realized completely.

The effective alteration of our environment is accomplished through technology. The contents of a technology are moulded by our changing ideas of well-being. Technologists do not concern themselves with the quality of the chosen idea of well-being. They consider it their task to find technologies adequate for the realization of any idea of individual well-being or the well-being of groups, irrespective of any coherent or long-range consideration. Technology is the study of the means by which nature is transformed into items useful for our own material life. Technology is thus often considered merely a means towards an end. The means, in turn, are looked on as things by themselves. But means arise through our intentionality, and intentionality points to our power of conferring meaning. Modern physics, for instance, confers meaning by tracing the phenomena of nature to their invariant causes. With Helmholtz we make force the cause of natural events, and nature in the context of modern physics is a calculable coherence of forces.

Modern technology is not merely a means towards an end but itself partakes in the constitution of our world. Modern technology requires that we constitute the world as the one objective world. Objective space is continuous and homogeneous, and it contains the bodily things. On the basis of this conception of space, we are able to construct a framework of Cartesian coordinates, external to the objects. By means of such a coordinate system, we are able to define the location of any object; and in terms of the object's potential movement, we can define the direction as well as the speed of the movement.

Mathematical reason creates space that exists by itself. This space may contain bodies whose position is not determined by the nature of the bodies themselves.[10] The bodies are essentially all alike and they are normally in motion, so that their positions with reference to the Cartesian coordinate system are always changing. We need to measure with precision the changes in the positions of bodies and account for these changes in terms of the underlying physical forces. In this way, we construct a theory of nature based solely on bodily things.

According to Heidegger, the modern physical theory of nature prepares the way for modern technology.[11] Although modern technology is chronologically later than modern physical science, it is in its essence historically earlier. When Heidegger intends to elucidate the essence of modern technology, he wants to grasp not the real object, in this case technology, but the essence of technology. He states, "When we are seeking the essence of 'tree,' we have to become aware that what pervades every tree, as tree, is not itself a tree that can be encountered among all the other trees. Likewise, the essence of technology is by no means anything technological."[12]

Heidegger reminds us that the word "technology" is derived from the Greek word *technē*, which means "it reveals whatever does not bring itself forth" (such as a garden or a bridge).[13] But what is responsible for bringing something into appearance? There is *physis*, which is bringing something into appearance from out of itself – the wilderness brings itself forth. There is also *poiēsis*; its coming into being is not found in the being itself, for it requires that someone bring it forth – the silversmith brings forth the silver chalice. *Technē*, in turn, belongs to *poiēsis*, to bringing forth. The Greek word for truth is *a-lētheia* which means "to reveal, to bring into the open." Heidegger points out that *technē* is a way of revealing. According to modern technology, nature is fully revealed in terms of physical reason. Energy, which is concealed in nature, is unlocked. The kind of revealing that occurs in modern technology is "unlocking, transforming, storing, distributing, and switching about."[14]

In naturalism, truth is seen as the outer correctness of the object in its representation. In modern technology, the elements of nature lose even their status as object and attain a state of objectlessness in the form of what Heidegger calls "standing reserve." The particular manner of revealing in modern technology is based on our intention to obtain maximum yield at the least amount of expense. The object as object disappears. For example, the forest becomes a standing reserve – merely a potential in the form of building material for the building industry. A mountain becomes a potential for the extraction of minerals for the mining industry. The world is revealed as standing reserve, to be ordered when the sense of need arises. Agriculture becomes agro-industry and represents a "challenging forth" towards maximum yield and minimum expense. The great danger inherent in modern technology is referred to by Heidegger as "revealing which orders," which "challenges."[15]

Technology, which orders the self-revealing as standing reserve, is founded on "enframing," as Heidegger calls it: "[Enframing] is the way in which the real reveals itself as standing reserve"; and "enframing is the gathering together which belongs to that setting-upon which challenges man and puts him in position to reveal the real, in the mode of ordering, as standing reserve."[16] In Heidegger's view, the state of mind of en-framing blocks our ability to bring reality into the open, where we may learn about it. This blocking represents an extreme danger for us, for it keeps the reality all around us concealed from us, and thus we necessarily act blindly in relation to it.[17]

SUMMARY

The view came to prevail that there is not human mental life as such, but that there is solely a historical mentality. According to historians, con-sciousness is not fixed; it changes with history. The historian in turn has access to a particular historical mentality whenever he or she employs the appropriate method in the historical investigation.

In the contemporary interpretation of the relationship of human beings and nature, the objective view of the world has become all-pervasive. In order to feel secure among the things around us, we intend to create rational spaces out of nature in its natural state. The effective alteration of our environment is accomplished through modern technol-ogy. Modern technology reveals nature in the form of physical theory; it employs exact physical science, which entraps nature as a calculable coherence of forces. Things in their very nature constitute themselves from the process of technical production. Instead of appreciating things in their own right, we relegate them to the objectlessness of standing reserve.

### CHAPTER FIVE

# Nature and Culture, 1

## CULTURE IN TERMS OF WHAT
## HUMANS HAVE ADDED TO NATURE

Natural forms are forms that bring themselves forth out of themselves. This is in contrast to implemental things and works of art, which do not bring themselves forth but are produced by humans.[1] More broadly, we may want to distinguish between natural things and cultural things. Cultural forms are those that humans have created out of nature or have permanently added to it.[2] Agriculture is an example. As the term implies, agriculture is a cultural form. Objectively, it is a means of transforming the structure of an ecosystem (a unit of nature) into a system that favours the production of food for human use. Early in the nineteenth century, Albrecht Thaer spoke about "rational agriculture": "The rational study of agriculture must therefore demonstrate how the greatest amount of profit can be extracted from a farm."[3] According to the rational position, animals and plants as well as inanimate nature are material assets. Earth and land are a commodity. Valuation of land is linked to production.

Wild plants and animals have their natural places, but natural places are abundant only in natural areas. Wilderness indicates the abundance of special places for these plants and animals. The state of the wilderness therefore secures this abundance. While natural landscapes imply plenteousness of wildlife, rational landscapes in the service of humans carry, as a consequence of human action, the scarcity of special places for both wild plants and wild animals. As the result of agricultural activity, species diversity declines – as has been the case for many years. The philosopher of culture Ludwig Klages published an essay in 1913, "Man and the Earth," in which he lamented the loss of animal life-forms.[4] The threat to species viability lies primarily with the loss of habitat, and loss of habitat is associated with the exploitation of natural environments.

The conservationist has taken the level of diversity of plant and animal life within a given area as a measure of the natural order, and species extinction as a measure of the destructive influence of humans on that order. We can distinguish between natural extinction and extinction caused by humans. Norman Myers points out that extinction brought about by humans is not a deliberate act of elimination, carried out by a few people, but that it occurs through the destruction of habitats. The destruction results from the activity of a larger number of people in the pursuit of the "consumerist life style," which is described, in its broadest sense, as striving for maximum yield at minimum expense.[5]

One speaks also of the natural environment and its modification through human influence. This area of inquiry is known as "environmental science" and deals with the problems associated with increasing human population, increasing use of resources, and pollution.[6] For example, modern agriculture uses large amounts of inorganic fertilizer. Some of this input of nitrogen and phosphorus nutrients runs off from the soil into lakes and streams, and leaches into groundwater. This is a major cause of eutrophication – the high productivity of a body of water due to nutrient enrichment. Under high productivity in a body of water there is a vast increase in the growth of minute life-forms called plankton. Ultimately, these organisms die and sink towards the bottom and are decomposed. Their decomposition uses up oxygen, which is required by fish; and when the water is deoxygenated, it can no longer support the life of fish. Eutrophication as a natural process is extremely slow, but the natural process has been accelerated as a result of the runoff of nitrogen and phosphorus nutrients. Because of inputs of pollutants resulting from human activity, the main body of a lake deteriorates in quality at a rate much greater than that of normal aging.[7]

There appears, then, to be a direct conflict between the existence of natural things and cultural things. Environmental science has made it its task to inquire in general into the modification of the natural environment through human influence. It adopts impact studies in preparation for action in relation to the environment. In general, any act may be viewed as consisting of three phases: the phase before the execution of the act, the execution, and the phase following the execution of the act.[8] The first phase is a phase of planning. It requires our interpretation of environmental reality. The act is to be carried out in the real world, but reality, as we have seen, is to some measure hidden from us. For the act to be effective, we anticipate the real world; we devise a theory, a mental construct of aspects of reality. In the context of the world of science, only the existence of material bodily things is to be acknowledged, and the only relations acknowledged between them are those of cause and effect. The knowledge of the objective world, as we pursue it in science, is considered to approximate reality as it really is. The knowledge of the objective world must therefore form the foundation of the planning phase preceding the execution of the act.

Accordingly, while scientific theories form the basis for the planning phase preceding the act, the decision to act represents the transition from the first to the second phase. It is a confrontation between the world of anticipation and the real world. The one must be brought into line with the other, so that intentions and plans turn out to be realistic. Any gap that exists between the situation as viewed by means of the planning theory in the context of the objective world and the reality of the situation at hand represents a danger for the success of the act.[9] Action implies the realization of ends on the basis of the knowledge and application of appropriate means. Action, such as that of a farmer in relation to a field, implies the realization of ends through the alteration of the existing environment. Finally, the completion of the act allows the acting person to grasp the nature of the act as a whole in a reflective manner.

Environmental science holds that the various parts of the ecosystem and their relations to one another can be specified objectively. The knowledge of these relations allows us to predict what will happen to the ecosystem when a particular factor is changed. In his book *Environmentalism*, T. O'Riordan describes one mode of viewing the environment and acting in relation to it as the "technocentric" mode.[10] This mode is based on the notion that the natural environment consists of objects that need to be explored because they may be useful to our material life. This requires an objective rational assessment of natural resources as well as their management; it requires that we assess the value of the natural environment in terms of its natural resources and estimate the costs of the alterations imposed on it as a result of human intervention.

## THE RATIONAL ORDER AND THE
## ESTRANGEMENT OF THE ENVIRONMENT

The conservationist wants to preserve the natural order with its abundance of special places for both plants and animals, and it is the state of the wilderness that secures this abundance. According to rational principles, on the other hand, a space is to be structured in such a way as to secure maximum yield at the minimum expense, for human material consumption.

Nature in the state of wilderness is to be replaced with a structure that is governed by the *logos*, our thinking turned into reality – rational spaces. Rational knowledge is to advance civilization into a standardized and uniform factual world. To illustrate this, let us again take agriculture as an example. According to rational principles, agricultural production must be made independent of locality. The principal source of sugar used to be sugar cane grass, a plant of the humid tropical lowlands. For their sugar supply, the lands of northern Europe used to be dependent on the importation of sugar from such regions as the West Indies. Disruption of the transport route at various times, as a result of political events, and the will towards self-sufficiency in the supply of sugar led to the cultivation of the sugar beet plant. This plant belongs to the temperate climatic zone, and consequently the conditions prevailing in northern Europe were suitable for its agricultural production. But because of the heavy weight of this root crop, effective commercial use had to wait until the building of a railway network. As a result of these introductions, the agricultural production of sugar has today become largely independent of location.

Rational agriculture pursues the aim of maximum crop production. The principal limiting factors for the growth of plants are the supplies of nutrients. In traditional agricultural practices, nutrients were supplied by spreading manure on the field. The amount of manure available depended on the number of animals on the farm, but additional growth could be achieved by applying artificial manures such as guano from Chile. Through chemical technology, Fritz Haber and Karl Bosch achieved a method whereby, under high temperature conditions, high pressure, and in the presence of special catalysts, nitrogen and hydrogen from the air could be chemically combined to form an artificial inorganic nitrogen fertilizer. Thus, by the removal of the limiting conditions, optimal crop production is achievable.

Environment, as "that which environs," refers to persons and is bound to location. But such a notion is unacceptable to our rational viewpoint, which acknowledges the presence of bodies in uniform space only; so environment also has to be made into an object that can be studied scientifically, as in environmental science. In his essay "The Morality of Music," Rudolf Kassner suggested that people have created for themselves a mirror in which they see their dreams come alive, and that for these people reason is such a mirror.[11] Reason, through the use of concepts, gives things a definitive form, which according to Kassner covers and hides the content in the same way that a mask covers and hides the face. While we may have access to the nature of a person through that person's face, the mask prevents us from gaining access to it, and so the world becomes inaccessible to us – the world remains hidden from us. In *The Natural Alien: Humankind and Environment*, Neil Evernden views humans as "natural aliens," meaning that by their very nature they seem unable to make sense of their place within a wider context.[12] Our rational manner of dealing with "what is" objectively has moulded a world that is becoming progressively more monotonous. As a result of this prevailing trend, we lose our sense of place to which we belong and even feel alienated from our "environment."

We seem not to be able to understand the environment in its relation to ourselves. Erwin Straus calls our experience the "I-Allon" relation.[13] He defines Allon as the organized wholeness of the surrounding composition within which the person moves and acts. Literally, *allon* is a Greek word meaning "the other." The Allon is a hidden reality, deep and distant. We do not know at any particular moment what the prospects are for the I-Allon relationship. We may even become fearful because we do not know how we stand in relation to the Allon. Each one of us is in search of the meaning the Allon possesses for us.

Let us take a brief look at the I-Allon relationship, as it arose in twentieth-century Expressionism, as one face of art. According to Walter Falk, any event taking place at the interface of a person, as a member of a particular cultural group, and aspects of that person's circumstance consists of at least three components: the motive power, the power of resistance, and a third component, the outcome of the event resulting from the meeting of these two powers.[14] In Expressionism, the motive power is a demonlike god whose character is made up of destructive drives. Opposite him as the power of resistance stands the living space of human beings, such as the city. Using the city as example, Falk refers to Georg Heym's poem "The God of the City."[15] The city tries to resist the demonic god's will to destroy it by offering him the respect of all its citizens. But this only incites the god to display his true nature by destroying the city that venerates him. The structure of the event expresses a profound manner in which the world may be experienced. Our will to penetrate the earth in the manner of "enframing," in order to maximize the products that we can extract from the earth for our own material life, has become unrestrained. This will thereby becomes transformed into destructive power, the demonic god of Expressionism.

Expressionism acknowledges the existence of eternal powers, the presence of powerful pregiven orders beyond the individual which penetrate a person's existence. We were sure that we make our own images, but now we realize that in fact we do not. Images are not things that are isolated; images are beyond the individual.

### THE SITUATION CIRCLE

Culture, it will be recalled, has been defined as what we have made out of nature or what we have permanently added to it. We have attempted to replace nature with a rational world, and thus we could say that the world arising from our reason is culture. But we always find ourselves in a situation. Each individual constructs his or her environment in the form of an environmental situation. According to the psychiatrist Thure von Uexküll, we interpret the environment initially in terms of a problematic situation, in which the assignment of meaning is not immediately followed by a fixed pattern of behaviour. A problematic situation leaves open various possibilities of action. The range of possibilities is first envisaged in the form of images. The various programs of action, which are either inborn or acquired, are checked for their applicability against these images. Only when the structure of a program is thought to be appropriate for the solution of the problem at hand does action follow. Thure von Uexküll describes this sequence of events in the form of what he refers to as "situation circle"[16] (figure 1). According to the situation circle, being healthy implies being able to construct a situation that allows a solution to the problem at hand. Failure to construct such a rewarding situation leads to sickness.

Culture is evidently not a readily composed structure founded within our reason, because our own well-being is inextricably bound up in the environmental situation. The importance of the composition of our surrounding space for our well-being, especially in reference to mental health, is pointed out to us in an article concerning schizophrenic patients: "There is only one thing we can say today about the influence of

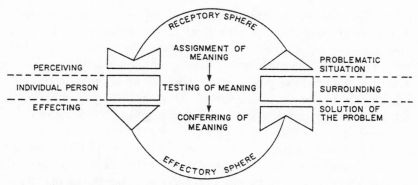

Fig. 1 Situation Circle (adapted and modified from Thure von Uexküll, "Der Situationskreis," *Lehrbuch der psychosomatischen Medizin*, Urban & Schwarzenberg, 1981)

the surrounding with certainty: people suffering from schizophrenia react in a particularly negative manner to strong obtrusive emotions in their environment. They need a restful, friendly and trustworthy 'climate.' They require a surrounding which keeps them from sinking into apathy through tasks which they are able to solve and duties they can carry out. They must have room for making decisions which bring them back into reality. Any surrounding which does not fulfil these two requirements – emotional protection and encouragement to do things – makes the course of the disease turn to the worse, be it the clinic, the family, a home or a resident community."[17] The particular composition of the surrounding space apparently both enhances and constrains us. The psychologist Eugène Minkowski has described how one of his schizophrenic patients[18] perceived space: this person constantly related external events solely to himself, thus showing the tendency for conglomeration in lived space or a deficiency of lived distance.[18]

The importance of the family as part of the patient's environment is stressed by psychiatrists R.D. Laing and A. Esterson. In their book *Sanity, Madness and the Family*, they analyse a family whose daughter spent nine of her last ten years in a London hospital. The daughter states that "blackness came over her when she was eight."[19] The same authors describe another family whose daughter suffered from schizophrenia. The parents apparently aggravated their daughter's plight. The mother says, " 'It's not much good trying to get people who are already in a tight corner themselves, is it, that's the point'."[20] Space, as the patient's mother experiences it, appears to have profoundly affected the life-space of the daughter.

The structure of the psyche is fragile and may, under certain conditions, become severely distorted. Minkowski describes a patient, a talented painter, who began to hear conversations and experienced various bodily sensations accompanied by these voices.[21] Each movement of his hand while painting was attended by these conversations. The voices, or auditory hallucinations, came to superimpose themselves on the normal perception of reality in his everyday life. Two lived spaces, one superimposed on the other, existed in him simultaneously, a light space and a dark space. In this patient, the light space and the dark space were not separated in time like day and night but were experienced simultaneously. The pathological reality had no exact location; it fully engulfed this person. Minkowski speaks of an absence of the fullness of life. By fullness he means "Life which unfolds around us and in which we take part has fullness."[22]

In the fullness of life we acknowledge coincidence and chance, while in the pathological world of the schizophrenic person events are of a predetermined nature. In the fullness of life we experience space as being socialized, in contrast to the desocialized space in the experience of the schizophrenic. In contrast to the crowding in the lived space of the schizophrenic or the deficiency of lived distance, the fullness of life is characterized by free space in which the person's actions can readily unfold.

SUMMARY

Nature is that which brings itself forth out of itself. It provides a multitude of special places for both wild plants and wild animals and hence assures great diversity of life-forms. Culture is composed of forms that humans have created out of nature or have permanently added to it. Taking agriculture as an example of cultural form, we realize that it greatly diminishes the diversity of special places for both wild plants and wild animals. As a result of agricultural activity, therefore, species diversity declines. This is especially true of rational agriculture, which aims at maximum yield at minimum expense. Such rational practices require the conception of a uniform space in which locality as a factor in agricultural production is eliminated, as are other factors that limit the growth of plants and animals. This conception and the use of rational spaces may, however, have unforeseen consequences, such as the eutrophication of bodies of water following the application of inorganic fertilizers in the production of crops.

Our aim has been to replace nature with a rational world. Since we have defined culture to be what we have made out of nature or what we have permanently added to it, the rational world is synonymous with culture. We find ourselves, however, in a situation that is not solely of our own making. A situation is a constellation of which we form a part. The rational approach, giving permanent meaning to the items composing the environment (as well as their utilitarian interpretation), leads to our own estrangement from what surrounds us. We have to construct our environment in the form of an environmental situation. We view it in terms of a problematic situation. Being healthy implies, according to Thure von Uexküll's situation circle, being able to construct an environmental situation that allows a solution to the problem at hand.

### CHAPTER SIX

# The Biological World

## THE SURROUNDING WORLDS OF ANIMALS

In the objective world, we acknowledge a reality that is composed of objects. These objects are linked insofar as they move in relation to one another. The linkage is one of cause and effect as the forces of nature cause the events. Like Helmholtz, we make force the cause of natural events. In the one objective world there are no subjects and there is no meaning, because we acknowledge only cause-and-effect relations.

Let us now look at the nature of the relationship between an animal and its environment. The interpretation of this relationship in terms of the stimulus-response model was discussed earlier. According to that model, the stimulus as a physical entity causes a fixed response of the organism. But is the relationship between the organism and its environment solely of a causative nature, the stimulus being the cause and the organism's response the effect? Might not the constitution of the environment be perceived directly by the animal in terms of its "affordances" (what the environment affords the animal for its life)? In his theory of affordances, the eminent student of vision James J. Gibson proposed an affirmative answer to this question.[1] In his theory of affordance, Gibson, distinguishes substance and medium as being two aspects of the environment. The animal moves about within the medium – the fish in the water, the bird in the air. Substances are material bodily structures and are important for animals because they may mean food, as seaweed does for the sea urchin, or they may even represent poison, as the foliage of the European yew tree does for cattle. The interface between the substance and the medium in turn represents a surface, such as the surface of seaweed in the medium of seawater. The surface may be important for the animal: fish spawn their eggs over beds of seaweed because the surface of the seaweed provides for the attachment of the fish's eggs and for their support in the water. Gibson thus understands affordance to be a particular combination of properties of substance and surface in relation to the animal. Affordance signifies what the environment provides for the animal, such as the flat surface of the ground which enables the animal to stand on it; it is, as Gibson expresses it, "stand-on-able."

On this basis, Gibson makes a clear distinction between "habitat" and "niche." The habitat is the space in which the animal lives, while the niche is a set of affordances that the animal uses. "Competence" means the ability of an animal to exploit fully the set of affordances offered by an environmental situation. The animal perceives this situation while it acts. The environment offers many affordances, and an animal fits a niche when the niche provides the right set of affordances for that animal. Gibson emphasizes that affordances are real, not in terms of the physical level, such as matter and energy, but at the level of meaning. An animal acts in relation to things according to the significance they hold for it within the whole setting.

According to Gibson, the animal directly perceives the meaning of things and beings in their setting in the animal's surrounding space. For the animal, the meaning arises from the affordance the environment offers it. An animal's most immediate experience of the environment lies in its grasp of the layout of the space in which it acts, as it does in self-movement. But does this laying hold of the environment occur through perceptual grasping only, for instance, through visual perception of the environmental situation? It has been shown that passively transported kittens, for example, learned nothing visually about the terrain: purely visual teaching of the layout was ineffective in terms of improving the kittens' performance. The surrounding space of the animal is perceptible to it only through its own actions.

Does not each animal, then, possess a core that correlates sensing and acting? And are not sensing and acting, in turn, linked to certain aspects of the animal's surrounding realm in terms of the meaning these aspects have for the animal's own life? In 1928 the biologist Jakob von Uexküll published his *Theoretische Biologie*, in which he revised the principles of biology. According to Uexküll, biology deals with the principles of the autonomy of life: the world is not a one-level reality, because there exist

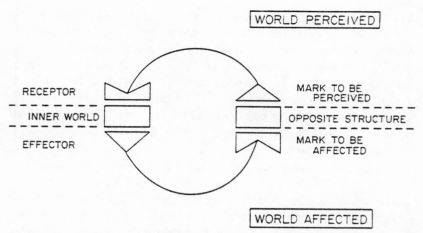

Fig. 2   Functional Circle (adapted and modified from Jakob von Uexküll, "Der Funktionskreis," *Theoretische Biologie*, 1928)

characteristic biological phenomena beyond the physical ones. Uexküll understands biology to be the study of the surrounding worlds of animals. An animal's behaviour and the nature of that animal's surrounding space are determined by the animal's plan of construction. This plan relies on the parts composing each animal, their structural quality, and how they combine into one integrated unit.[2] Is the animal, then, not really a machine, similar in its nature perhaps to a steam engine? An animal is similar to a machine insofar as a machine also is made up of parts that are put together to make a functioning unit. But a major difference is that a machine does not integrate its own surroundings; it is part of the surrounding space of humans. A machine is a creation of human beings, created by them in order to serve them.

Each animal has its own particular kind of surrounding space, of which it forms the centre. Things and beings entering this space take on particular meaning according to the kind of animal. Its surrounding world is composed of a perceived world and a world to be affected. Subject and object are not separated by a dividing line. A part of the organism, its receptors, and a part of the object, the side of it that is to be perceived, are linked. Similarly, the effectory organs and that aspect of the object to be affected are intertwined. This can be represented in terms of what Uexküll calls the "functional circle"[3] (figure 2). According to this scheme, perceived and affected worlds are interconnected to form a unit. The object participates in the action because it has characteristics that serve as bearers of marks to be perceived and also as bearers of marks to be affected, both of which together represent the "opposite structure." Within the "inner world," the objects of the surrounding space are represented in the form of schemata. Depending on the species of animal and hence its plan of structural organization, schemata may be kept quite general and may combine a variety of objects under the same category of meaning, or they may be very exclusive. The inner world forms an image of the particular surrounding world, giving it meaning and guiding actions in relation to it.

A simple animal has a correspondingly simple surrounding world, and a complex animal has a more complex one. The tick hangs motionless on a branch at the edge of an opening in the forest. Through its position, it can fall on a passing mammal. In the diagram in figure 2, we would place the tick into the functional circle as the subject, with its own inner world, and would represent the mammal as the opposite structure. In the absence of the mammal, there is no mark to be perceived by the tick. But when a mammal approaches, providing the tick with the opportunity to obtain some of the animal's blood (which the tick needs for maintaining itself and for producing offspring), something miraculous happens: of all the characteristics of the mammal that might be received by the tick, only three are perceived. These succeed one another in a particular sequence and lead the tick securely to its goal. Of these three functional circles, the skin gland of the mammal represents the first mark, because the stimulus of sweat triggers within the tick's receptor a specific mark to be perceived. Processes corresponding to the schemata in the tick's inner world link the receptors to the effectors. Certain impulses in the effectors make the tick's legs flex, allowing it to fall from the branch of the tree onto the hairs of the mammal's fur. The hairs are the second mark to be perceived in the form of touch, and this erases the previous mark, the mammal's sweat. The new mark triggers a search by the tick until it finds its way to the warmth of the skin, the third mark. Once it attains contact, the piercing of the skin begins. Of the many distinctive features that the mammal exhibits to us, only three become marks to be perceived in the world of the tick.

Uexküll therefore introduces the subject into the study of biology. Animals are inseparably bound to their surrounding world. Each animal perceives and acts in relation to the things and beings in its whole setting according to the meaning these possess for the animal. The positivists maintain the lawfulness of the objects themselves, their unchangeableness, but Uexküll holds that there is not one objective world; there are as many subjective worlds as there are different kinds of animal:

Whoever has dealt even to only a limited extent with the surroundings of animals will never think to ascribe to objects their own characteristics as such which make them independent of the subject. The changeability of the object is the most important law. Each being changes profoundly when it is brought into a different surrounding. The floral stalk which bears in our surrounding a flower, becomes in the surrounding of a grasshopper a tube filled with liquid, out of which it pumps the liquid which it needs to build its house composed of foam.

The same floral stalk becomes for the ant a gangway directed upwards connecting its home-nest with its hunting ground. For the cow grazing upon the meadow, the floral stalk becomes but a small part of the tasteful fodder which it chews and pushes into its wide mouth.[4]

Hence, for Uexküll, objects are constituted by the subject, and it is the changeability of the objects that needs to be stressed. One object occupying a particular space within the framework of one objective world constructed in thought can have very different meanings within the context of different surrounding worlds, which may overlap with one another quite loosely or more densely.

The organism performs in an integrated manner in relation to the meaning things and beings have for it when they enter its surrounding world. The contents of the surrounding to which the organism is coassigned are present also in its inner world in the form of schemata, or archetypes. Uexküll gives as an example the relationship between the spider and the fly. The architectural design of the spider's web is based on the characteristics of the fly, as if the web represented "a subtle painting of the fly which the spider traces upon its web."[5] In the development of the fly there exists what Uexküll calls an "archescore" of the fly, and in the development of the spider there exists an archescore of the spider; but the archescore of the fly has an effect on the archescore of the spider, making the spider project the archetype of the fly into its own surrounding space in the construction of its web.

The study of surrounding worlds takes as its starting point the assumption that the whole of living nature represents a total interconnection. The whole is not determined by its parts; rather, the parts are determined by the whole. The all-encompassing plan of the whole is like a vast network of interconnections of meanings. This network is not visible as such, but by studying the surrounding worlds of organisms, we are able to gain access to this perspective of the biological world. We should note, however, that Uexküll maintains that we are not able to know how the surrounding world of any creature appears to that creature. We can merely study the structure of its surrounding world by studying the structural plan of that organism, the types of receptors and effectors it possesses, and how the organism performs in relation to the various things and beings in its surrounding world.

In addition to a physical reality, we recognize the autonomy of life-forms within the context of a biological world. In this case, the study of surrounding worlds still takes, as the basis of its approach, the outer appearance of the organism in terms of the organism's structural plan; but it goes beyond outer appearance in that it attempts to gain access to the meaning that things and beings are given by the organism as subject. Uexküll penetrates beyond the opaque surfaces of the outer appearance of the organism into their invisible contexture.

### THE NATURE OF SIGNS

When we view the organism as physiological body, we explain its behaviour on the basis of the mechanics of the body. This applies to all living beings, including plants. Many plants, for instance, produce nectar, a sweet liquid, and release it by special structures known as nectaries. These are found on flowers (floral nectaries) and on vegetative parts of the plant (extrafloral nectaries). The latter are found on leaves, as in passifloraceous plants, and on the leaf petioles of various species of plum, as well as on other parts of some plants. For example, the rubber tree *Hevea brasiliensis* has compound leaves, and where the three leaflets join there are three large nectaries, the nectar of which fuses into one large droplet. When examined at the end of the night following its secretion, the composition of the nectar as well as the concentration of the sugars it contains are very similar to the solution found in the food-conducting elements of the plant. A tree with 10,000 young leaves secretes in this manner 50 grams of sugar per day.

According to A. Frey-Wyssling, even though the growth of the leaves has been completed, the translocation of the food in the food-conducting elements continues. Since the sugar is no longer used by the leaves, it must be eliminated through the nectaries, which have therefore been referred to as "sap valves."[6] One can think of a plant as a form of container. Large parts of it are constructed as tubular hollow spaces containing fluids. When the pressure in these tubes becomes too great because of the forces underlying translocation, the "sap valves" release some of the fluid and thus also release the pressure in the plant's fluid-containing spaces.

In an article entitled "Ant-Plant Mutualism: Evolution and Energy," Brian Hocking refers to the significance of extrafloral nectaries and other structures in terms of mutual relations between animals and plants as well as between different kinds of animals.[7] For example, while various plants shelter ants and offer them food, such as the nectar secreted through extrafloral nectaries, the ants can play a major role in sheltering the plants:

It is difficult to understand how anybody who has encountered an ant-acacia in the field in East Africa could doubt the efficacy of the ants in protecting the trees. Sometimes even when a tree is approached and always as soon as it is touched, the ants pour out of their holes in the swellings and scramble toward the ends of the branches, emitting an odor which most people find repulsive, while adopting an attitude with the abdomen erect which even a formicophile can recognize at once as threatening. It has been argued that this is in defence of the brood; but how can the brood be protected without protecting the tree which protects and feeds it? The prompt aggressive response to any kind of interference is apparently characteristic of mutualistic plant ants.[8]

In addition to nectaries, many plants that have mutualistic associations with ants produce special nutritive structures, rich in other nutrients. These structures, which may be spheroidal and distinctly coloured, sometimes also house the ants. For instance, *Crematogaster* ants inhabit stipular swellings of *Acacia drepanolobium* trees. These ants are major items of prey for mantids, and there is a striking resemblance between the mantids and the stipular swellings of the acacia tree. Consequently, the ants may regard the mantids as a home to be sought rather than recognizing them as an enemy to be avoided. An animal may thus need to avoid a certain condition of its surrounding world while seeking out another specific condition. The organism evidently gives meaning to things and beings in its surroundings and acts in accordance with the meaning.

In such a relationship there needs to be an interpreter. When we think about objects in the naturalistic context, they are nothing but objects. Any changes of the object in relation to the environment are fully accounted for in terms of physical principles, such as the workings of the nectary in terms of a pressure valve. But the object also stands in relation to someone, such as the acacia tree with the *Crematogaster* ants. The object as significant form is brought forward by the interpreter. How is this made possible?

The organism as interpreter gives meaning to things and beings in its surrounding world, and the meaning reveals itself in the form of signs. The aspect of the object that brings about an effect in the interpreter is known as the interpretant. A sign is therefore something that stands for something else and has meaning for someone. According to Charles S. Peirce, there are three aspects of signs: the sign itself, the sign in its relation to the object, and the interpretant that is necessary for the functioning of a sign.[9] The interpretant is what a sign produces in the interpreter; for example, a particular act. In the relationship between mountain-ash berries and the birds that eat them, the redness and round shape of the berries serve as the sign linking the object (berry) to the interpreter (bird) through the interpretant. The power that resides in the sign may give the bird a habit of action, because the meaning that the sign has for the bird in terms of the affordances present in its surrounding world is fixed through the schemata or archetypes present in the inner world of the bird.

Something functions as sign when it is placed in a triadic relationship – one where it stands for something else (the object) and where what is signified has meaning for someone. The triadic relationship involves a sign process, or "semiosis." The sign brings about a reaction in the interpreter, and the aspect of the sign, the interpretant, is subject to convention. Thus, habits are the final interpretants of signs. The interpretant of a sign is also defined as "a disposition to act in a certain way."[10] Charles Morris represents semiosis as a five-term relationship – $v$, $w$, $x$, $y$, $z$ – and illustrates this by means of the dance of the bees, as discovered by zoologist Karl von Frisch.[11] A bee discovers a source of nectar and is able to direct other bees to it by returning to the hive and performing a dance in front of them. The dance is the sign ($v$), and the other bees are the interpreters of the sign ($w$). The disposition inherent in the bees to react to the dance in a certain way is the interpretant ($x$). The kind of object in relation to which the bees are capable of acting in this fashion is the significance of the sign ($y$). The $z$'s are the contexts in which the sign occurs. In the present example, the position of the hive is part of the context.

Let us look once more at the trees that bear nectaries. Frey-Wyssling considers the tree by itself – as a physiological body. The presence of the extrafloral nectaries are accounted for in terms of the operational principles underlying the mechanical functioning of the tree, that is, in terms of a "sap valve." But the tree may also be considered as a figure in relation to the ground in which it is embedded. In this case, the interface of the figure and the ground comes to the fore. This interface emerges in the form of perceptual meaning, and meaning reveals itself in the form of signs.

The sign stands in a relationship for the object to the interpreter and therefore forms a link between the object and the interpreter. The ant, for instance, may be the interpreter, and the object may be the swelling found at various sites on the acacia tree which serve the ant as a dwelling. The dwelling is represented in brightly coloured form, which is the sign linking the object to the interpreter. This linkage establishes the natural form of the swelling as the intentional form in the world of the ant. Similarly, as Brian Hocking reported, when one touches one of these acacia trees or even approaches it, the ants leave their houses and rush to the tips of the branches, producing there a repulsive odour and taking on a threatening posture. These are signs that link the object, "space of tree," to the interpreter in such a manner as to bring about a habit of action in those who are touching or merely approaching the tree, preventing them from interfering with that particular space, which is the space of the mutualistic relations between the ants and the tree.

The study of signs is known as "semiotics," a term that originally referred to medical symptomatology. Symptomatology is "the study of symptoms; that branch of pathology which treats of the symptoms of disease; also, a discourse or treatise on symptoms."[12] "Symptom," as it applies to human pathology, is defined as "a (bodily or mental) phenomenon, circumstance, or change of condition arising from and accompanying a disease or affection, and constituting an indication or evidence of it, a characteristic sign of some particular disease."[13] In this case, the disease as the object has many different aspects. It is the choice of the appropriate sign that gives us access to the true nature of the disease. We may choose as the sign the reading on a thermometer measuring the temperature of the body, and this temperature we then take as a symptom representing the disease. Clearly, a sign is not the condition itself; it is the representation of the condition. The interpretant, as one aspect of the sign, can bring forth an answer in the interpreter of the sign. One may, then, broaden the notion of symptom beyond its medical use so that it means the sign of a condition in general. Natural phenomena, too, can be viewed in terms of symptoms and can be interpreted as signs.

For the knower, there are not objects as such. The "being" of the object depends on its being represented in the form of a sign. We need to interpret a world, but inherent in interpretation is the possibility that we interpret erroneously. In this regard, it is important to note that Peirce distinguishes between the "dynamic interpretant" and the "final interpretant."[14] The dynamic interpretant is the relation that is brought about in the interpretant by a sign. The final interpretant is the interpretation that the interpreter is supposed to attain if the sign is fully considered.

CAUSATION VERSUS SIGNIFICATION

It will be recalled that Uexküll describes the relationship between the organism and its environment in terms of the "functional circle." The functional circle in turn serves as a model for semiosis. Within the organism's inner world, the objects of the surrounding world are represented in the form of schemata. Through the perceptors, the inner world forms an image of the particular surrounding world, giving it meaning and guiding actions in relation to it. In the study of the biological world, the notion of meaning is unavoidable. It involves subjects for whom something has meaning. The order of biological reality consists of invisible spheres of surrounding worlds of organisms, which are interconnected through meaning (for instance, the world of the predator and its prey). Furthermore, it is this meaning that shapes the diverse life-forms in networks of interconnections. The painter August Macke has given an excellent example of an interconnection: "The flower opens at twilight. The panther stoops, the prey in view, and its powers grow through this sight; and the tension of its power determines the length of the jump."[15] According to Uexküll, the whole of nature is interconnected according to plan through invisible strings of meaning. It is this whole structure that Uexküll calls "nature."

Nature is evidently not governed solely by physical forces. In addition to the reality based on physical reason, there is the biological world founded on vital reason. We must acknowledge that the structure of nature is based on the presence of different forms of relations. Charles S. Peirce identifies three classes of relations: monadic, dyadic, and triadic.[16] Each of these three classes is irreducible, so that triadic relations cannot be reduced to dyadic or monadic relations. Monadic refers to mere appearance – for instance, a colour – regardless of whether it is perceived or not. A complete triadic relation is one in which no two of the three correlates can be correlated to one another without mediation by the third correlate. As we have seen, this applies to the sign relationship because three aspects must be present: the object, the sign, and the interpretant. Here, the interpretant serves as the third correlate, and only with it does the relation become triadic, that is, significative. This is distinct from a dyadic relation, in which one thing acts upon another – for example, the environment in the form of stimuli acts upon the organism, which is made of irritable matter, and brings about the response. Causal relations are based on one thing acting on another; they are relations in terms of "brute action."[17] Causation is a dyadic relation. As soon as an interpretant is present, the relationship becomes significative because of its triadic nature and it becomes distinguished from a causative relationship, which is dyadic. We must therefore distinguish between causation and signification. It is due to the interpretant that a situation is triadic and significative, in contrast to a causative situation, which is dyadic because of the absence of the interpretant.

According to Uexküll, we must make a clear distinction between the inorganic world and the organic world.[18] The inorganic world is ruled by monadic and dyadic relations (the latter being relations in terms of "brute action," one thing acting on another) whereas the organic world is governed by triadic relations. The relationships that participate in the bringing forth of natural things are therefore also very different in inanimate and animate nature. In the inanimate realm we are dealing with causative relations, while in animate nature it is the significative relations that predominate.

Semiosis, the sign process, therefore refers to the influences between the sign, the object, and the interpretant.[19] A good example of the relationship between the sign and the object is weather. Since weather as an object has many different facets, we must choose a particular sign to represent the object weather – for example, the barometer reading that measures changes in atmospheric pressure. The pressure of the atmosphere is but one aspect of the total object weather. Nevertheless, these two are directly linked. Such a direct linkage between the sign and the object has been termed the "index" by Peirce.[20] There can also be a relationship between the object and the sign in which the two are not linked. The sign merely resembles the object. Peirce cites a map as an example of this type of sign. In this case, the sign may possess its particular character by virtue of its own internal nature only and without regard to whether the object actually exists or does not exist. Peirce gives the name "icon" to this class of signs."[21]

In addition to icons and indices there is a third class of signs, which Peirce calls "symbols." As he explains, "There is, however, a third totally different order of signs, which become such, not by virtue of any character of their own as things, nor by virtue of any real connection with their objects, but simply by virtue of being represented to be signs ... Such signs may have little or much internal meaning and external meaning but they have a third kind of meaning which consists in the character of the interpretant signs which they determine. This is their principal meaning."[22] For Peirce, the meaning of this class of signs is that they prescribe an effect in the mind of the interpreter. In this regard, symbols represent objects according to convention, which is usually a structure of general ideas. This in turn is a "habit"; symbols result in the establishment of habits of action.

Beyond the practical skills of animals, human beings have developed a symbolic imagination. Our ability to construct different worlds in thought is made possible through the intervention of our symbolic forms. By means of our symbolic gifts, we are capable of drawing attention to something that is not given as such. We are thus not fixed in the present within a surrounding world, as the animal is, but are potentially open to the world. The forms that the objects of nature have in our thought are cognitive forms as distinct from natural forms. The cognitive forms must be consistent within the architectonic structure of knowing as a whole. This structure is dependent not only on the nature of our sensory experience but also on the intervention of our symbolic forms.

SUMMARY

Can we sustain the view of nature as the one objective world, as it is composed of bodily objects and as the relations between them are governed by causation? Can we maintain the intellectual perception of the organism-to-environment relationship in terms of the stimulus-response model? These are questions that we have begun to address here. Because we acknowledge in the one objective world only bodily objects and cause-and-effect relations, subjects and meaning do not exist. Concerning nature, we need to look more closely at the organic world and especially at the life of animals. In contrast to the theory of one objective world, Jakob von Uexküll speaks about the surrounding worlds of animals. In Uexküll's functional circle, the animal as subject gives meaning to things and beings in its environment and acts accordingly in relation to it. Meaning in turn reveals itself in the form of signs, and this implies a relationship in the presence of an interpreter. Signs link the object to the interpreter. According to the significance the objects have for the animal, as given to it in the form of signs, each kind of animal has a characteristic surrounding world.

The subject interprets the world through signs that represent the real world. A distinction must therefore be made between causation and signification. Signs imply significative relations because of the presence of the interpreter. In the absence of an interpreter, relations are causative, are based on one thing acting on another thing. Nature is then no longer an objective space containing bodies: in the presence of the organic world, nature represents a network of interconnections of meanings in terms of the various surrounding worlds.

### CHAPTER SEVEN

# The Life-World, 1

## THE BODY SUBJECT AND BODIES IN SPACE

While Jakob von Uexküll speaks of an animal's surrounding world, Martin Buber questions whether it is appropriate to call the surroundings of an animal a "world."[1] He recommends that we use the term "realm" and reserve the concept "world" for humans. Buber argues that because of the symbolic nature of human beings, it is characteristic of them to have to build a world. Humans do not have a world that is fixed for their kind; instead, they are capable of creating different worlds. The recognition of this basic difference between animals and humans justifies Buber's critique of Jakob von Uexküll's term "surrounding world" as applied to animals. We need to speak about an animal's "surrounding" or "realm." Only humans potentially have a world, because they make it with the things.

But we must recognize that all living beings are not merely bodies in space. What characterizes organisms is the nature of their boundary layer. We may say that the surface of an inanimate body forms an interface between the body and the medium – the boundary is where the inanimate body stops, and it is identical with the contour of that body. By contrast, an organism contains a boundary layer that forms part of the body, setting the boundary layer distinctly apart from the medium as well as from the body itself.[2] This makes it possible for the organism to be a subject, which is directed both out beyond the body that it is, and back into it again. Thus, the core of the body is capable of detecting aspects of the body's environment.

Do we merely perceive objects in our environment? The philosopher-anthropologist Helmuth Plessner asks what the distinguishing features of living things as perceived objects are, in contrast to non-living things.[3] He compares our perception of a tennis ball with our perception of a cat. The cat confronts us; it takes its own place. It has not only a position in the Cartesian coordinate system but it has positionality, it has its natural place. The tennis ball is merely a body in space, defined by its position in the coordinate system. Science conceives of the human body, too, as being part of a bodily world that is independent of the experiencing subject. But we possess a dual knowledge of the human body, formed out of two very different experiences. We know the human body from the outside, and we also have an inner perception of our body. Everything manifests itself to us through the perception of our inner space.[4]

In our natural attitude we find ourselves in the world of our everyday experience. The life-world is the world of our experience – a reality in which we partake. The space we inhabit is not objective space; instead, there is a vital experience of space. This experience involves an act of perceptual integration. Perceiving, knowing, and skilful acting arise from a centre within each person. This centre is the "body subject," as the phenomenologist Merleau-Ponty calls it.[5] In everyday life, the whole individual being attends to things as they are in their settings, when perceiving them, knowing them, or acting in a skilful manner in relation to them. But what do we mean when we speak of the body subject? We are used to thinking in terms of the duality of mind and body, and we usually do not use the term "body subject," for it fits neither the notion of body nor that of the mind. We need to distinguish between the body subject and the physiological body. While the physiological body is defined by what is present inside the body's outer contour, the body subject reaches out beyond its outer boundary into the world. It acts intentionally and intelligently, but precognitively.

The vital functions of a body subject are carried out within a vital space, which is not given as such but is shaped by that being's actions. Let us think of the hunter who shapes his intentions within the surrounding wilderness. During the hunt the animal being hunted must first be detected. The hunter is in a state of complete alertness, which involves the whole body subject intently looking at everything in sight and yet not focusing on any particular element of the total wilderness composition in which the hunter is embedded. "The hunter's soul leaps out, spreads out over the hunting ground like a net anchored here and there with the fingernails of his attention."[6]

The hunt is made possible because the hunter is congruous with the hunted. But the hunt represents a task in which the hunter makes the moves towards his intended goal, the death of the hunted animal. The hunt is a game, because the outcome of the hunt is indeterminate, like the indeterminateness of the outcome of any game – the game of chess, for instance. The hunter's will to realize the image of the successful hunt shapes his every move and overcomes every obstacle within the landscape. The hunter's objective study of the conduct of the hunted animal enables him to predict the nature of the animal's evasiveness. In the hunter there is also a sense of pleasure at the beauty of the healthy animal and the way it relates itself to the surrounding wilderness. Finally, the hunter senses the sacredness of the innocent animal that is to be sacrificed in the hunt. Ortega y Gasset speaks of the mystical union formed during the hunt – the hunter experiences the space of the environment "from the point of view of the prey, without abandoning his own point of view."[7]

## THE GESTALT CIRCLE

Spatial constellations are very different when thought of in the context of objective space or when experienced in everyday life, because experience involves an act of perceptual integration. Here we should consider the notion of gestalt. "Gestalt" means that we primarily perceive an entire configuration rather than its individual parts. The whole is in some manner before and beyond its parts: the way we interpret the parts is influenced by the function they assume within the whole. We perceive landscape, for example, as gestalt, for the parts are grasped simultaneously in their relationship to one another in the form of gestalt.

Here, a distinction must be made between a deterministic world as it is constituted in science and a life-world that is the world of our experience.[8] The life-world is the region of our ordinary experience, out of which all meanings must emerge. The self gives meaning to its circumstance and acts in relation to it. Acts involve both sensing and moving; perception of our surroundings and our movement in relation to it are linked. There are certain accommodations that function to keep the body in contact with particular aspects of the surrounding world until these contacts are actively interfered with. We conduct ourselves in time through our movement in such a way that perceptual coherence with those particular aspects is maintained. The form of movement arises out of the development of the meeting ground of a person and selected aspects of that person's surrounding world. Self-movement is an act representing the genesis of form.

The psychiatrist Victor von Weizsäcker stresses the primacy of form over time in self-movement. He measures, for example, the time it takes to draw large and small circles in the air, and notes within wide limits a constant "figure-time"; that is, a large circle is drawn with a much greater linear velocity than a small circle. Organic movement takes place in such a way as to assure the completion of a figure of performance in the same period of time, irrespective of the size of the figure. Movement at its very beginning already anticipates the figure of the performance: the movement from its inception takes into account the total performance of its figure and its completion, and hence sets its own time at the start.[9]

All motor performances are sensed by the self. We attend to things and beings in our surrounding space by being aware of changes within our body. Hence, an immediate contact is made between the self and the composition of the space within which the self moves. In his book *Der Gestaltkreis* (The Gestalt Circle), Weizsäcker examines the relationship between perception and voluntary movement.[10] The essence of the gestalt circle is that perception and movement are fungible, that one may be used in place of the other in the expected and intended attainment of the goal. Perception is seen as a meeting of the self and aspects of its circumstance. In the case of visual perception, the eye extends the body into the surroundings and forms a particular unity with them. Perception linked with movement is always an active development of the meeting ground of self and aspects of the surrounding world – a process in which the goal is anticipated but cannot be foreseen. Because perceptory qualities participate in this gestalt circle, we can no longer maintain a totally separate treatment of the quantitative objective world and the qualitative subjective world, for in the unity of the self and its surrounding, genesis of the form of movement arises in order to maintain coherence between the two.

Psychologists have demonstrated that in the visual perception of two stationary objects placed and presented successively, the objects may be grasped in the form of one gestalt: what was presented earlier and what is presented later are brought together into one unitary event; the movement of but one object is recognized. When separate points of light are presented in darkness, perception establishes mutual relations of movement between them. One point may objectively be at rest, the other be in motion, but perception distributes velocity and direction of movement to both points in some definite relationship of one to the other. When normal observers are presented with a configuration of points of light dispersed in an otherwise dark background, they construe the spatial organization in a way not consistent with Euclidean geometry.

The seeing eye establishes lawful relationships between the movements of the light spots, a phenomenon that Weizsäcker calls the nomophily of perception;[11] "nomophily" is the tendency to create lawful relationships between perceptory events. Perception of gestalt takes time when viewed by an observer who is outside the subject who perceives the gestalt, because it requires both calling to memory, "anamnesis," and anticipatory performance, "prolepsis," as Weizsäcker calls it.[12] Gestalt requires concurrence of events in time, or synchronism of what is objectively no longer present and what is objectively not yet present. Gestalt, according to Weizsäcker, is primary and time is secondary, because it is the gestalt that participates in the structuring of biological time. Biological time is interpreted as the present that bridges time. Time develops within the gestalt, and the gestalt circle is the genesis of the form of movement.

Every perception and movement is initiated by a meeting of a person with aspects of that person's surroundings. Such a meeting on a meeting ground is realized because it is a possibility and not a necessity. To illustrate how a person meets something, Weizsäcker gives the example of a chess game.[13] The game is created through the decisive moves of the two players. The goal of each is to achieve a contexture of the two sets of figures in which the opponent's king cannot escape. Thus, the goal or purpose of each player's performance during the game is to defeat the other player, and each player moves in anticipation of the other player's countermove. The two sets of figures of the two players represent an extension of themselves; both players attend to the changing configuration of their opponent's set of figures in relation to their own. Thus, both players, from their own set of figures, attend to their opponent's set of figures within the dynamic relationship of the whole. In a game of chess, each player considers the other player to be the opponent in the context of a contest, and each seeks self-realization. The realization of the whole game is bound to the players' adherence to the rules of the game and their freedom in each move. The reality of the game is brought about because it contains the scope of the indeterminateness of the nature of the countermove.

Having thus described the nature of the game of chess and the role of the players in creating the game, Weizsäcker stresses the difference between a scientist and a game player. The scientist is a person who, as impartial observer, asks about cause and effect within the determinateness of the one objective world. Weizsäcker compares the scientist with a person who knows the rules of a game but does not create it. But according to Weizsäcker, science ought not merely to explain phenomena; it ought to create a reality in the same manner as the chess players create the reality of their game of chess.

## THE GESTALT COURSE

The body subject is grounded in the earth, but the earth is concealed. Through the body subject's perception and movement, "gestalten" come out from concealment into view. Various linkages thus arise between the gestalten in our surroundings and the way we act in relation to them. The philosopher and art historian Hermann Schmitz speaks of "gestalt courses" – tendencies towards forms of movement suggested to the body subject through the gestalt.[14] Schmitz distinguishes between the aspects of our sense of feeling that can be localized and those that cannot. An example of the former is our sense of feeling tired. By contrast, feelings of being cheerful or sad cannot be localized; they are spread over the whole expanse of experience. Consequently, Schmitz attributes them to the soul, in contrast to the sense of being tired, which can be localized within the body.

Schmitz distinguishes also between two kinds of localities: relative localities and absolute localities. A relative locality can be identified by its position and distance from other localities. Schmitz refers to the relative localities in their relation to one another as spatial orientation. In contrast to this type of locality, the state of being tired is localized in the body subject's sensing and therefore has an absolute locality. The locality of the sense of being tired does not have to be ascertained by means of spatial organization, for it is present in the body itself. The locality of the body is thus an absolute locality and represents a whole that cannot be subdivided into elements. In addition to the body subject's disposition as a whole, the body subject contains various islets, such as the sense of hunger and thirst, which are more localized within the body. However, the body subject's absolute locality as a whole encompasses the islets, and the locality of the body subject's sensing as a whole is therefore an absolute locality. The localities of the body, such as its component organs, are not sensed by the body subject itself; they are given to us merely as relative localities. Similarly, the localities of the soul, such as sadness, are not delimited but are spread over the whole field of experience. Only the phenomena of the body subject are characterized by absolute localization. Thus, the body subject's sensing of its own warmth is a state of being warm, which is localized absolutely and is distinguished from the thermal condition of an object, which is localized relatively while the body subject touches it.

The state of the body subject is characterized by a number of different categories, such as contraction and expansion. Contraction, as it may occur in fear or pain, leads to tension. By contrast, expansion leads to inflation, as when pride takes over. The mutually competing effects of tension and inflation may occur simultaneously or sequentially, one alternately dominating the other. When they are simultaneous, the body subject senses intensely, but when tension and inflation occur sequentially, we obtain a sense of rhythm. A further category characterizing the state of the body subject is its sense of direction. Schmitz envisages direction as the sense of being harnessed into the irreversible gradient between the tightness of the body itself and the wideness into which it proceeds. This sense is rooted in the absolute locality, because the spatial form of the present is the narrowness of the body subject moving over into the wideness that is spatially indeterminate. Finally, the movement of the body subject is pervaded by the contrasting tendencies of sensing either a sharp pointed mark or one that emanates in a faint or diffuse manner. These tendencies Schmitz calls "epicritical" and "protopathic."[15] Epicritical refers to the tendency to locate places distinctly; protopathic is the tendency of locations and contours to become indistinct.

Gestalt suggests to us a certain form of movement. While we are intensively grasping a particular gestalt, there arises within us a suggestion for movement that is characteristic for that type of gestalt. This suggestion of the form of movement, coming into being on the basis of the gestalt, is the gestalt course belonging to it. We need to trace the gestalt course back to the gestalt and to characterize the gestalt by means of joining it to modes of sensing. Schmitz discusses, for example, the protopathic gestalt course of a rounded form, such as the figure of a half-sphere. This figure may be viewed either in the form of its convex curvature or its concave curvature. With the convex figure, the curvature bends away from the viewer, and the viewer is thus placed at varying distances from the various points on the curvature of the surface of the half-sphere. Different viewpoints are therefore required in order to get an equally good apprehension of the various segments of the convex half-sphere, but this cannot be attained all at once. This type of gestalt draws what is kept in sight into the depth, and the mode of looking takes a course towards the depth and away from the viewer.

If the gestalt shows itself in its concave aspect, then, at least from a certain standpoint, all points of the curvature are at about the same distance from the viewer. Instead of keeping the viewer at a distance, the gestalt enwraps the viewer. Furthermore, in the pursuit of the rounded concave curvature, the movement of whatever one is looking towards finds its completion when the image, such as the image of a cupola, absorbs it. The gestalt from its concave aspect invites the viewer to stand right in its midst, while the gestalt from its convex aspect encloses itself against and away from the viewer. In terms of the arousal of movements, radial spikes emanating from the centre of the cupola bring the concave half-sphere into firm relationship with the body subject's sense of contraction and expansion – with the body subject's sense of tension and swelling, and sense of direction.

In contrast to the profound arousals of movement initiated by gestalten that are curved, a straight line offers practically no such suggestions. For this reason, we have a tendency to seek out figures in which a number of straight lines intersect, creating corners. These figures, in terms of their gestalt courses, give rise to epicritical tendencies in ourselves, as body subjects, and make us find distinct localities.

In support of his thesis about an affinity of gestalten and the body subject in terms of the arousal of forms of movement, Schmitz cites the case of a physician's patient. Because of a pathological condition, this patient, even though able to see, was unable to recognize even the simplest gestalt characteristics, such as a straight or curved line. He was able, however, to recognize these characteristics if he actually traced the features of the gestalt. So although his ability to recognize the movement of the gestalt, as suggested by the gestalt directly, seems to have been lost, the patient was able to grasp the movement suggested by means of self-movement, which made up for what the eye failed to do. We are reminded here of Weizäcker's gestalt circle, the essence of which, as noted earlier, is that perception and movement are fungible.

A stationary gestalt may thus stir us to carry out a particular form of movement – it may give rise to a gestalt course of ourself, as body subject. Self-movement is an act that represents the genesis of form. In order to carry out such an act, we have to surrender ourselves, as body subjects, to the impact that a stationary gestalt has upon us, or we have to design a field of movement in our imagination. Within the form of expressive self-movement, the body subject may wish to portray aspects of the life-world, such as the experience of a path or a slope of a meadow. Expressive self-movement enables us to express what we wish to portray.

Spatial gestalt is not merely the summation of elements but a whole unitary complex of the imagination. This is the basis for the idea of the "quality of gestalt."[16] Each gestalt has a certain level of configuration, and we are able to compare the quality of different gestalten. We can characterize a gestalt by the product of the unity and diversity realized within it. The greater the product is within a gestalt, the higher is the level of its configuration. Thus, a gestalt attains a higher level when, with the same degree of diversity of its component parts, a more rigorous unity is achieved. Where there is the same degree of unity, the gestalt attains a higher level when it encompasses greater diversity.

In the life-world there are intimate connections between a body subject and the gestalt of the surroundings. As mentioned above, Schmitz speaks of our sense of swelling and contrasts it to our sense of contracting. He describes what the body subject senses when entering something very large and impressive: "We experience a sense of expansion in all directions, with a sense of liberation drawing a deep breath when we enter a high forest or an unexpectedly beautiful hall. We inevitably expand our chest, and make ourselves bigger, just as if we wanted to adapt ourselves to the imposing surrounding and prove ourselves in this manner and by stretching out we have the vital experience of conquering the space and of expanding our sense of potency."[17]

Thus, a gestalt such as a landscape brings forth gestalt courses. As noted above, the curved half-sphere, which has both a convex and concave aspect, can give rise to very different movements from these two different views. In the painting by S. Haase depicting a hillside and its reflection in the water (figure 3), the hillside is the convex aspect and its reflection is the concave aspect. The convex hillside retreats into the depth of the space of the landscape, but the reflection of the hill, because of its concave form, draws our view forward to embrace us and make us part of it. Much the same point was made by Prince Pückler-Muskau, who, although he wrote in the first half of the nineteenth century, has had a lasting influence on the art of landscape design. He decided to transform Muskau, his estate, turning it into a landscape park. In the de-

Fig. 3    S. Haase, *Landscape* (Kraft von Maltzahn, Halifax, Nova Scotia)

sign, he considered the contours of the land to be of great importance. Muskau is a valley, a concavely curved structure, and Pückler pointed out that this characteristic gives us a sense of inwardness. But he added that we also have a desire for the distant landscape, as we experience it from the hilltop or convex structure, which gives us a sense of outwardness.

## SUMMARY

Nature includes the organic world. In its presence, nature is composed of an intricate complex of what Jakob von Uexküll calls "surrounding worlds." Martin Buber recommends that we reserve the notion of "world" for human beings and use the term "realm" when referring to the surroundings of animals. Humans have developed a symbolic imagination, and by means of their symbols they are capable of building different worlds. While the animal's surrounding is fixed for its kind, a human being has the task of building a world.

We cannot experience the objective world as such because it is a construction in thought. The world of experience, on the other hand, is a life-world. Experience and construction in thought are therefore different, although one affects the other. The instrument of experience is our whole body subject, which is different from the physiological body. The physiological body belongs to the objective world and its dimensions are defined by the body's outer boundary, whereas the body subject reaches out beyond its boundary into the world; it acts intentionally and intelligently but precognitively in relation to the world.

Perception sets up contact between a body subject and aspects of its surrounding world. The body subject gives meaning to these aspects and acts in relation to them. Acts involve both sensing and moving; perception of what surrounds us and our movement in relation to it are linked. We are connected with our circumstance and there are certain accommodations that function to keep our body subject in contact with particular aspects of the surrounding world. Various linkages arise between gestalten in our surrounding space and the way we act in relation to them – our gestalt course. The gestalt course is a set of movements that come into being in the body subject on the basis of a particular kind of gestalt.

### CHAPTER EIGHT

# The Life-World, 2

## MODES OF SPATIAL EXPERIENCE

In our natural attitude we find ourselves in the world of our everyday experience. We participate in this world, our life-world, by seeing and moving. We usually perceive our surroundings when we are moving, rather than when we are standing still. The movement may not even involve the whole body; it may concern only parts of the body, such as the eyes, the head, or the trunk. When we are moving actively in relation to what surrounds us, the surrounding also moves; the surfaces of objects close to us move faster than those farther away from us.

The term "surfaces of objects" is used rather than "objects," because we perceive the surfaces of objects against a continuous background surface. Since surfaces reflect light, James Gibson speaks of the "ambient array," meaning the structure of the light reflected from objects in an environment.[1] The structure of the array, in turn, is determined by the properties of the object. Since the normal person has two eyes and since each eye perceives one array at any given place and time, two arrays are present. When we move, we perceive a sequence of arrays. The sequences remain constant over time, and Gibson refers to them as "invariants of arrays."[2] Because of their constant nature, the invariants make it possible for us to gain a visual understanding of characteristics of the environment, including those of objects that form part of it. It is these sequences of the arrays that define the invariants of the environment.

When we move in the environment, the perception of motion in the frontal plane is determined by the aiming point of the movement. The point towards which we move is the centre of the radial outflow of the ambient arrays.[3] The flow of the arrays can be represented in a motion picture, and even though in this case our own position is fixed, in that we are merely observers, we experience a sense of movement towards the centre of the outflow. This situation is another example of Weizsäcker's gestalt circle (that perception and movement are fungible). Visually, the movement of our own body and the movement of our surrounding are as one. In the previous chapter, when discussing the sensuous aspects of the experience of gestalten, such as building structures or landscapes, we noted that, according to Schmitz, a stationary gestalt suggests to us, as body subjects, certain movements. The gestalt and the body subject form a temporary union, which unfolds through the quality of time, always starting from the present, now, and proceeding towards the future.[4]

The psychologist Erwin W. Straus has focused on the nature of space in the individual's immediate experience.[5] He contrasts this space with Euclidean space, which represents the conception of space in the one objective world constructed in thought and which underlies the conceptual system of physiology. Straus notes that in personal experience there are "gnostic" and "pathic" modes.[6] These two modes do not participate to an equal extent in the various kinds of perceptual experience, such as touching, hearing, and seeing. One may look at something and experience it as an object situated in the space opposite oneself and separate from oneself. Hence, when one is "looking at," the gnostic mode dominates one's experience and develops what one is looking at in its characteristics as object. In seeing, the objects stand out from their ground, sharply demarcated in terms of their contours and with their colours remaining attached to them at their surface. The gnostic moment emphasizes the question, "What is it?" – leading to our preoccupation with the conceptual grasping of the world.

In contrast to the gnostic moment, the pathic moment of experience puts emphasis on the preconceptual communication with the phenomenally given, the changing manner in which things appear directly to our senses. While the gnostic moment dominates our experience in the sense of seeing, the pathic moment rules our perception in touching. Straus elaborates on the pathic moment of the experience of space within the acoustical sphere. Whereas colour remains attached to the object, tone separates itself from its source, approaching us and filling the space around us; whereas we behave actively and directionally in relation to coloured objects, tone pursues us. Thus, optical space is dominated by direction and distance, but the spatial structure brought about by music is governed by symbolic qualities. The pathic moment does not ask, "What is it?" but "How is it?"

According to Straus, a fundamental difference therefore exists between the spatiality produced by colour and sound, for colour does not separate itself from the object, whereas sound does separate itself from its source. Yet the distinction between the experience of space in the optical and acoustical spheres may not be this rigid, as the following incident illustrates. Entering his studio one evening, the painter Wassily Kandinsky noticed a painting of great beauty that he did not remember ever having seen before. But after close examination, he suddenly realized that it was one of his own paintings, standing on its side. Because he had not recognized any objects in the picture by themselves, the colours had separated from the objects, as had the forms, and they had become abstracted, giving rise to a primarily pathic mode of experience of coloured space.[7]

Music is temporally differentiated, and music thereby differentiates time, and there is thus a direct link between hearing and movement. According to Straus, optical space gives rise to directional movement through that space, starting from one point in order to reach another point. Acoustical space is the space of the dance, and within this space we move without direction. In dancing, we experience our own bodily existence in relation to the surrounding space. In the movement of the dance, the body abandons its usual rigid vertical orientation. The body space so experienced is expanded into the surrounding space, filling it in all directions. When the trunk itself is placed in the movement, as happens in the dance, the live body space opens up in the widest manner possible and the dancer's experience is a mode of being truly present in the world, of fully participating in it. For example, in a trance dance, the use of touch is a way of communicating this sense of contact with the phenomenal world.[8] Dance space is not part of directional, historical space but is a symbolic region of the world. In this connection, Straus mentions the myth of Orpheus in which "men and animals, trees, forests, and even rocks, mountains and streams followed his sounding lyre."[9]

The above discussion focuses on perceptory experience in the acoustical sphere, contrasting it with the experience in the optical sphere, but usually various spheres are present simultaneously in the environment. The whole sensory envelope creates in us the sense of spatiality according to which we act. Because all motor performances are sensed by an individual, an immediate contact is brought about between a person and the composition of the space.

## FUNCTIONAL FORCES

In our life-world we must shape coherences between ourselves and aspects of our surrounding (the genesis of form of movement is the need to maintain the coherence between the two). The aspects of the composition of the environment with which we cohere depend on our intentions. Our acts are intentional, and intentionality confers meaning on the composition of the space in which we act. All areas in which human actions are involved are configurations, because configurations are based on both the functional forces arising within the individual person (or group) and the structure of that person's life-space. The functional forces are the "modes of a subject's self-realization vis-a-vis the external world."[10]

There are not merely individual people; but there are types of people who share a set of common traits, such as a set of functional forces. Spranger speaks of human life-forms – the basic forms of human life that are realized in people.[11] One may delineate the type of character, for instance, by the natural tendency of people to act in a certain way in a given set of circumstances. Spranger distinguishes between economic, aesthetic, religious, and other life-forms. Accordingly, a person belonging to an economic life-form will experience a particular set of circumstances differently from a person belonging to the aesthetic life-form.

When a man who is a farmer ploughs his field, his image of the future in relation to the field is the harvest. The plough is his implement, shaped to serve him effectively in the attainment of the anticipated goal. The functional forces coming into being in the farmer and dominating his acts are practical forces, and he gives his attention to the space of the field, for the field is to be created by him according to his image of it. But even though practical biological and economic forces may dominate the manner in which the farmer views his field, this does not necessarily imply the exclusion of other functional forces, such as the aesthetic sense or ethical and religious attitudes. The farmer must recognize the quality of ploughing, the manner of ploughing in relation to the nature of the soil, the time of ploughing in relation to season and weather, and the direction of ploughing as determined by the layout of the land; but the quality of the act depends also on the farmer's skills, which are in turn related to the attitude he adopts, the standards he sets for himself.

### PLACE-BALLET

The farmer's acts in relation to the space of the field he ploughs are based largely on what geographer David Seamon calls "body-ballet," a set of integrated movements sustaining a particular task or aim. Similar to body-ballet, a "time-space routine" is a set of habitual bodily behaviours that extend through a considerable period of time. The time-space routine as a whole has a certain pattern that unfolds. Sizable portions of a person's day – for example, getting up, going to school, and so on – can be organized around these routines. Because the body subject in its everyday life exhibits such preconscious intelligence in the form of skilful performances, it is free to devote its attention to other tasks.[12]

The farmer's experience of the world while working and his shaping of his world (the world of his field) are not the result of his being a looker-on who reasons theoretically but are the result of his skilful actions. These skills express themselves in our everyday life in body-ballets and time-space routines. They are preconceptual skills of our whole body as it relates itself to aspects of our surrounding world. We know that perception and movement are linked, but they are not necessarily voluntary; they may be induced through gestalten all around us. When gestalten evoke in us graceful expressions, the composition in which we are embedded and of which we form a part, such as a landscape, becomes a place *for* us rather than mere space facing us. We experience a sense of home, with everything that entails, such as the sense of belonging. A place is therefore a surrounding space which, in its whole composition, complements our own inward dimension while we are in unison with it.

Space compositions fuse body-ballets and time-space routines into a larger whole, which Seamon calls "place-ballet." Within such a composition, people following the pattern of their own everyday lives come together in space, which is now experienced in the form of place. Individuals begin to participate within the same space and create a larger place, with its own integrated pattern of life. As an example, Seamon describes the outdoor market in Varberg, a coastal town in Sweden. For hundreds of years, the market has been taking place twice weekly on an open cobblestone square. All the people at the market, including the vendors, obtain a sense of place, and this sense, as well as the site of the market itself, joins them together, shaping the whole into a dynamic rhythm.[13] But this happens only when the necessary spatial conditions, the appropriate number of people, and their purposes come together. Lewis Mumford, writing about the medieval town, has described some of the consequences of this type of supportive environment:

This daily education of the senses is the elemental groundwork of all higher forms of education: when it exists in daily life, a community may spare itself the burden of arranging courses in art appreciation. Where such an environment is lacking, even the purely rational and signific processes are half starved: verbal mastery cannot make up for sensory malnutrition. If this is a key, as Mme Montessori discovered, to the first stages of a child's education, it continues to be true even at a later period: the city has a more constant effect than the formal school. Life flourishes in this dilation of the senses: without it, the beat of the pulse is slower, the tone of the muscles is lower, the posture lacks confidence, the finer discriminations of eye and touch are lacking, perhaps the will-to-live itself is defeated. To starve the eye, the ear, the skin, is just as much to court death as to withhold food from the stomach. Though diet was often meagre in the Middle Ages, though the religious often imposed abstentions upon themselves in fasts and penances, even the most ascetic could not wholly close his eyes to beauty: the town itself was an omnipresent work of art; and the very clothes of its citizens on festive days were like a flower garden in bloom.[14]

## SUMMARY

The life-world is the world of our everyday experience, and we often experience our surrounding space while we are moving. Our self-movement makes our surrounding space move too. Visually, the movement of our body and the movement of our surrounding are as one. Erwin Straus distinguishes two different modes of experiencing our surrounding space: gnostic and pathic modes. In the gnostic mode, the conceptual understanding of the world dominates, while in the pathic mode emphasis is placed on preconceptual communication with what is given phenomenally. In seeing, the gnostic mode dominates; in hearing or touching, the pathic mode is the principal form of experience.

The mode of our relationship with our circumstance may therefore be very different, depending on which of our sensory envelopes is predominately actuated, for example, the visual or the auditory. How we act in relation to our environment is determined not only by the structure of the surrounding space but also by the nature of the various functional forces that arise within us and enable us to realize ourselves in relation to the world. Patterns of joint action may occur when people with related motives come together within a supportive environment. This pattern of shared action is called a place-ballet.

### CHAPTER NINE

# Visual Thinking and Tacit Knowing

## RATIONAL VERSUS RATIOMORPHIC PERFORMANCE

As mentioned earlier, in the everyday life-world the body subject is linked by invisible strings to its environment. Animals take their place within appropriate environments; they have positionality. But humans have placed themselves against their animal existence and have taken on what Plessner calls "eccentric positionality," meaning the attainment of distance from one's own body.[1] We are reminded here of humankind's affectations, the drifting apart of the seat of the soul and the body's centre of gravity when the human being becomes self-aware and makes nature into an object.

If we are to attain an objective view of the world, the subjective world must be eliminated. Nature has to be separated altogether from human life. With regard to the objective world and the subject's role in the construction of that world, John Locke introduced a distinction between the primary and secondary qualities of the object.[2] He called what the mind perceives "idea"; and he called the ability of an object to produce ideas the "quality" of the object. Hence, it is the quality of the object that empowers it to produce the idea of itself in us. Locke gave as an example the power that a snowball possesses. It is capable of producing in us ideas of white, cold, and round. The qualities are the powers of the snowball. Locke then discussed a grain of wheat that can be divided into two parts. Since each part retains the solidity, extension, figure, and mobility of the original grain, these characteristics are primary qualities of the object and produce simple ideas in us.

The primary qualities, such as bulk, figure, texture, and motion, cause our sensations. These sensations (colours, sounds, tastes, etc.) are not part of the object itself and thus are secondary qualities. For Locke, ideas arising from primary qualities are resemblances, while those of secondary qualities are not: "Take away the sensations of them; let not the eyes see light or colours, nor the ears hear sounds; let the palate not taste, nor the nose smell; and all colours, tastes, odours, and sounds, as they are such particular ideas, vanish and cease and are reduced to their causes, i.e., bulk, figure, and motions of parts."[3] After we have abstracted from the secondary qualities, it is the primary qualities that remain, and they give us direct access to the real world. Accordingly, if the objects of the outer world are to be established in the inner world as facts, this must be in the form of ideas based on the external object's primary qualities. Since we cannot experience the object's primary qualities as such, the objects of outer reality must be reduced from their appearance to symbolic formulation; they must be abstracted into their primary qualities.

This requires the construction in thought of a theory of the world. A theory is a system, structured through reason, that approximates a true description of the one objective world that is independent of the person. On the basis of the condemnation of the senses for the attainment of secure knowledge, as in Locke, we conclude that only rational performance gives us access to the real world. In the scientific analysis of the world, we are principally dealing with functional relations. General expression of functional relations is not possible by means of ordinary language but requires the symbolic structure of mathematics based on number. The abstraction of the objects into their primary qualities moves away from the need of vision for thought and leads to the autonomy of the mind and hence to "conceptual prediction of the unperceived."[4]

In the context of "critical philosophy," we think of knowledge as being fully explicit – statable in terms of language or its extensions, such as mathematics. Furthermore, language is considered to be the only means of thought. In language we must express our elementary ideas in sequential form, one after the other. Since the meaning given through language is brought together into a whole through the process of discourse, language is referred to as the "discursive" form of symbolism.[5] Only thoughts arranged in a sequential manner can be spoken, and they form the basis of rational thought.

Since the laws of nature to be discovered through science are concerned with the structural and sequential order of natural events, one needs to construct a system of discursive symbolism that corresponds to the necessary connections expressing that order. As has been noted, they are most readily articulated in terms of the appropriate mathematical forms, as an extension of language. Language largely contains and elaborates notions of attributes; and it arose within the context of experience, such as the experience of sacred space. Language did not come about through the construction in thought of the one objective world, and for that very reason it is not particularly useful for expressing the functional relations that science wishes to elucidate.

We consider only discursive symbolism – that is, language and its extensions – to be the bearer of our conception of the world. Accordingly, knowledge ranges only as far as it can be articulated discursively. However, the world presents itself to us in the form of signs. With Peirce we distinguish three different kinds of relationship between the sign and the object: a direct linkage called "index"; a relationship where the sign merely resembles the object – the "icon"; and, finally, the "symbol." This last class of signs determines an effect in the mind of the interpreter. In human life, language is a fundamental type of symbolic form. But it is not the only one.

Thomas Sebeok, one of the principal students of signs, contrasts human sign systems to those used by animals.[6] Those found only in humans he refers to as "anthroposemiotic." He distinguishes these from "zoosemiotic systems," which are those found in animals only as well as those found in both humans and animals. There is therefore no clear line where zoosemiotics turns into anthroposemiotics. Zoosemiotics is to a large extent based on both kinesics and morphology (meaning, respectively, communication by means of patterns of movements and expression through form). There are other zoosemiotic systems as well, such as signs in the disguise of smell. An example of anthroposemiotics is language, in which words and concepts are initially the outcome of very profound experiences and the reflective concentration of those experiences. Sebeok proposes that all sign systems not related to language should be considered to be zoosemiotic systems.

The knowledge of nature is, however, not limited to functional relations, which we express in mathematical symbolic forms based upon number; it extends to spatial relations, as in a picture. The picture is composed of elements, just as language is, but while the elements of language – the words and sentences – are understood sequentially, the elements of the picture are given simultaneously, and they are understood only in the context of the whole space of the picture. This space is purely visual (what the physicist calls "virtual space"), such as the space that is apparently behind a mirror; it is shaped for the eye only and is situated opposite the eye.

The meaning of the pictorial elements composing the picture is understood only through their relations within the whole structure represented in the picture. This form of symbolism is called "presentational" symbolism.[7] In addition to discursive symbolic form, meaning is conveyed by means of presentational symbolic form. The ethologist Konrad Lorenz distinguishes between two kinds of performance: rational performance and ratiomorphic performance.[8] In ratiomorphic performance simultaneity of forms dominates, whereas rational performance is based on discursive thought that can be spoken. An example of ratiomorphic performance is that of a medical doctor who discerns at a glance a certain complex of symptoms in the patient and suddenly recognizes the whole picture of the disease. According to Lorenz, the perception of such a picture of symptoms as a simultaneous phenomenon can combine a much greater number of elements, as well as their relations to one another, than occurs in rational performance.

Here we must return to the notion of gestalt. It will be recalled that gestalt means that we do not primarily perceive parts that we put together afterwards; rather, we perceive whole configurations. This being the case, I want to ask what the relationship is between perceptual experience, language, and thinking. This is an important question because we tend to believe that knowledge is based on verbal conceptual thinking alone. The argument goes approximately as follows: to grasp the general nature of something requires undertaking an act of abstraction, and abstraction, so it is argued, is a moving away from immediate experience. One appears to be dealing, therefore, with a very basic separation between thinking and perceptual experience. Accordingly, it is contended that abstract thinking is not based on sensory experience but takes place in the form of words. According to Herder and later to Cassirer, for instance, we are unable to attain an understanding of the nature of things by means of their visual characteristics except when they are associated with sounds. These sounds are in the form of words, and it is therefore only by means of language that a world emerges for us.

The plant morphologist Agnes Arber has suggested that the anatomist's notion of considering brain and eye to be distinct organs with distinct specialized functions has contributed significantly to the disjunction of mental thinking and vision.[9] Arber emphasizes that for

this reason many philosophers have belittled the visual aspects of thinking. She illustrates how we grasp the significance of a whole structure
through our ability to see. Our skill at comprehending intrinsic connections between the parts within the whole is necessary for visual thinking:

Every biologist must be able to confirm from his own experience that perception
depends upon preparedness of the mind, as well as on actual visual impressions.
As a trivial instance, the writer may recall having been acquainted with Queen-
Anne's lace (*Anthriscus sylvestris* Hofm.) for half a century, without noticing that
the pattern of its growth is such that the main axis almost invariably terminates
in a reduced inflorescence, which, in association with the grouping of the lateral
shoots below it, gives the plant a highly distinctive facies. When that visual fact
had at last succeeded in forcing its way into the mind, any plant that came under
observation was found to show this salient feature so strikingly as to leave the
observer bewildered and humiliated at having been totally blind to it year after
year.[10]

Evidently, there is not only mental thinking or verbal-linear-analytical
intelligence, but also visual thinking or visual-aesthetic-plastic apprehension. Knowing involves our bringing together an awareness of the elements of a composition with an awareness of their joint significance.

We need to turn to Rudolf Arnheim and his book *Visual Thinking,*
in which he intends to demonstrate that any kind of creative thinking
is a shaping of visual images.[11] Thinking requires as its object images
from the world of experience. Thinking has to do predominately with
the structural relations of things, not merely with matter or substance as
such. The visual sense is therefore the most important one for the act of
thinking, because seeing implies seeing relationships. Arnheim describes
the perception of form as the recognition of general structural characteristics. He points out that perception emerges as a means of recognizing
affordances in our surrounding space, and that this recognition is based
on categories, or qualities, rather than on individual incidents. An act of
selection is the most basic trait of perceiving in general and of seeing in
particular. This is achieved through a process of abstraction, because
only an abstraction can supply us with a particular principle of selection.
Seeing depicts kinds of things, types of objects and events. By "productive abstraction" Arnheim means the distinction we make between gen-
erative or central characteristics on the one hand and accidental or
peripheral ones on the other.[12] Knowing implies grasping the central
characteristics.

Our thinking cannot reach beyond the structures our senses supply us with. Images arise, of course, not solely in the present; they are potentially before us as images from memory. Arnheim points out also that the image of an object on the retina is formed not only by the physical thing itself but also by what surrounds it, and that the viewer is an important constituent of what surrounds the object. To see a thing means to see it in its connections. Mere naturalistic imitation does not really contribute to the interpretation of the nature of things. Perception is not a passive event in which we perceive something ready-made. Perception makes contact between a self and aspects of its circumstance. It is an active integration between the perceiver and the perceived. Spatial constellations are different when thought of in the context of one objective space or when experienced in everyday life, because experience involves an act of perceptual integration.

As mentioned above, abstraction involves an active selection, and it joins perception and thinking. Concepts and thinking are composed of the working up of images. Thinking applies concepts in the realm of perception itself as well as in the interplay between what we see directly and what springs up in our imagination. Perception and conception further one another because they are two aspects of a single unitary experience. A gesture, for example, has the power to impress us because it selects an essential characteristic of the topic that is being portrayed and accentuates it. But as Arnheim points out, the language of gestures demonstrates also that the perceptual qualities of form and movement are contained in the thought process itself.[13] These thought processes show themselves in the gestures; indeed, these qualities are themselves the medium in which thinking is taking place. And in the case of gestures, this is of course not restricted merely to visual characteristics, for the sense of muscular movement is an important aspect.

PHYSIOGNOMIES

It is through recognizing a gestalt that we have access to the significance something holds within a broader context. When we wish to gain access to the nature of another person, for example, we are able to do so by reading that person's face – gaining an impression of the person's physiognomy, or inner nature, as revealed outwardly. When we speak of physiognomy, we mean giving our attention to the exterior appearance of a person, the "superfecies," in order to know the character of that person. Physiognomy as a systematic field of study was founded by the Swiss writer and theologian Johann Caspar Lavater (1741–1801).[14] According to Lavater, physiognomy attempts to define interior characteristics from exterior traits. We judge what is not apparent, the inherent quality, through what is apparent to the senses, such as the movements of expression and action. We assume that a person's appearance, including movements, is the result of the interaction of that person's inner self and circumstance.

It will be recalled that chapter 1 described Alexander von Humboldt's explorations into the geography of plants. Humboldt applied the original notion of physiognomy, as referring only to human faces and characteristics, much more broadly to natural phenomena in general.[15] For him there was a physiognomy of vegetation, in terms of the "face" of an assemblage of plant life-forms – for instance, the aspects of a grassland or of a temperate deciduous forest. Furthermore, this physiognomy served as a sign, as we might say, revealing hidden features of a region.

A physiognomy is a form, and reading a physiognomy is a skill we have to learn. The recognition of the central characteristics of things and beings, as well as of assemblages, is based on experience. Experience is not something that we make anew together with the things; it is something that is already preorganized. The interpretation of a physiognomy is founded in our life-world, familiar images, and expressions. The manner in which we are open to the world, the way we are able to understand a physiognomy, and the mode of our experience of the world have already been formed into well-known patterns in which we feel at home. Nevertheless, new experience – for instance, the experience of an unfamiliar landscape – may also force us to reorganize our world. This reconstruction of our world may correspond more fully to what *is* and hence may result in the discovery of new patterns of meaning.

The philosopher-scientist Michael Polanyi views gestalt as the outcome of an active shaping of experience. Perception of gestalt implies that we grasp the whole aspect all at once as we comprehend at a glance the nature of a person through the appearance of that person's whole bodily expression. According to Polanyi, all knowing involves bringing a person's awareness of the elements of a composition together with an awareness of their joint significance for the composition of the whole. In the modern epoch, we have structured an ideal of detached knowledge of the world, concerned with the discovery of objective natural orders and laws based on value-free scientific rationality; we have relegated problems related to value to the personal sphere of feeling. For one objective world to be independent of the person, a split has been made between the subjective world and the objective world. In contrast to this position, Polanyi argues in favour of incorporating into our conception of knowledge the part that we ourselves necessarily contribute in shaping the knowledge. He believes we must abandon our current ideal of detached observation and replace it with a comprehension that all knowledge is necessarily personal knowledge. His main work is, in fact, entitled *Personal Knowledge.*[16] He gives his book the subtitle "Towards a Postcritical Philosophy," evidently in contrast to Descartes' "critical philosophy," which has dominated our thinking. Polanyi's theory of personal knowledge is based on the insight that all knowing involves grasping the relationship between a whole composition and the particular elements constituting it, as in gestalt. Polanyi speaks of a "from-to knowledge."[17] We focus our attention *to* the whole composition but *from* the particular elementary features. We are usually unable to specify the elementary features *from* which we attend *to* the comprehensive entity. We are not aware of the particular traits themselves, but we focus on their coherence; they represent clues to the reality of the whole composition.

Polanyi argues that one's skills are definite forms of one's knowledge. Skills require two different kinds of awareness of things: subsidiary awareness and focal awareness. The focus is generated by a performance, and the performance is guided by a purpose. The subsidiaries are related to the focus in a fundamental manner. Polanyi gives an example:

When I use a hammer to drive a nail, I attend to both, but quite differently. I *watch* the effects of my strokes on the nail as I wield the hammer. I do not feel that its handle has struck my palm but that its head has struck the nail. In another sense, of course, I am highly alert to the feeling in my palm and fingers holding the hammer. They guide my handling of it effectively, and the degree of attention that I give to the nail is given to these feelings to the same extent, but in a different way. The difference may be stated by saying that these feelings are not watched *in themselves* but that I watch something else by keeping aware of them. I know the feelings in the palm of my hand *by relying on them for attending to the hammer hitting the nail.* I may say that I have a *subsidiary* awareness of the feelings in my hand which is merged into my *focal awareness* of my driving the nail.[18]

We usually assume that all knowledge is explicit, with all its elements apparent. But physiognomies cannot be fully and clearly stated. There is a tacit component involved, inasmuch as we know more than we can tell. Polanyi speaks of the clues that are subsidiarily known as the "proximal term" of tacit knowing, and of the clues that are focally known as the "distal term" of tacit knowing.[19] Tacit knowing merges the proximal into the distal. In tacit knowing we attend *from* the proximal *to* the distal term. Such a subsidiary-from and focal-to relationship is maintained only as long as a person integrates the subsidiary to the focal through acts. Since these acts are acts of integration performed by persons, we must personally shape our knowledge; hence, according to Polanyi, all knowledge is personal knowledge, based on participation rather than on detached observation.

The knower may shift attention, though, from the focus to the subsidiaries. Consider the skill of pianists. Their focus must be on the music, and they must be only subsidiarily aware of the motions of their hands and of their fingers hitting the keys of the instrument. The individual tones receive a meaning only in the context of the whole musical composition. This meaning is wiped out when pianists focus their attention on the tone itself rather than on the music as a whole. This fixing of attention necessarily paralyses the pianist's performance. When we comprehend a particular set of items, such as the particular tones, as part of a whole, our attention is focused on the understanding of the joint meaning of the particulars, the gestalt of the music. All knowing is structured in this manner, and it is for this reason that Polanyi looks at gestalt as the outcome of an active shaping of experience. We perform this by means of our tacit powers. They always involve two items: attending *from* something *to* something else.

In Polanyi's view, the world is composed of strata of realities. These strata are put together meaningfully in pairs, each pair consisting of one higher stratum and one lower one. We cannot focus on the higher level until we have a tacit understanding of (or have put into subsidiary awareness) those things that must be incorporated into the higher level. Polanyi uses the example of speech, sentences being ways of expressing thought as well as being composed of vocabulary and grammar.[20] A child cannot learn to make sentences if it is still focusing on building an elementary vocabulary and accumulating the fundamentals of grammar. Only when we have dwelt in these aspects and when they become proximal, or an extension of the body, can they be attended *from* and directed *to* the process of sentence formation. Whereas Spranger maintained that we can dwell only in other persons, Polanyi proposes that all knowing involves a tacit component that requires our active personal participation in that aspect of reality with which we wish to establish contact by dwelling in it.

Polanyi discusses the relationship between our own body and the aspects of the surrounding space to which we attend: "There is one single thing in the world we normally know by relying on our awareness of it for attending to other things. Our own body is this unique thing. We attend to external objects by being subsidiarily aware of things happening within our own body. We may say to know something by relying on our awareness of it for attending to something else is to have the same kind of knowledge of it that we have of our body by living in it."[21] Each one of us perceives our body from inside, but healthy people are usually aware of their inner body merely in a subsidiary manner and reach out beyond their body.

The principles operating at a higher level cannot be derived from those governing the lower level. The organizational principle of a higher level exercises control on the particulars forming the lower level; this principle Polanyi calls the principle of marginal control.[22] A higher level can come into existence only through a process that does not exist at the lower level. The individual items of the lower level must therefore emerge into a coherent pattern which forms the next upper level. In moving to a higher level, however, the various possibilities for reaching above the boundaries of the previous level leave open the possibility of failure as well. Unless the higher level (for example, speaking in sentences) exercises a margin of control over the principles governing the lower level (in this case, vocabulary and the rudiments of grammar), the higher level will fail in its execution of that skill. The margin of control is the way in which the principles governing the lower levels are integrated in one particular combination for the successful achievement of the emergent principle. This means that in the example we are using, the speaker must choose the correct grammatical forms, such as the appropriate tense, and must also choose the right words so as not to suggest a wrong meaning. If these go astray, the sentences fail to perform their focus, which is to express a thought.

Perception is also a process in which we integrate clues into a focus. We focus on a coherent centre and thereby bring the subsidiaries into an order. In the perception of a physiognomy we give our focal attention to the face as a whole. In contrast to our focal attention, our subsidiary awareness is given over to the particulars forming the elements of the face. Since the proximal "from" is the subsidiary awareness, these particulars are joined together in the from-to knowledge of the physiognomy when we experience that face. Polanyi justifies his use of the terms "proximal" and "distal" by saying that all meaning tends to be displaced away from ourselves. Furthermore, the bringing into order of the subsidiaries by the focus requires that the person cross a gap which separates the clues from their organizing focus.

Perception of gestalt implies that we grasp the whole aspect all at once, as we comprehend at a glance the nature of a person through the appearance of that person's entire bodily expression. The performer coordinates his or her moves by dwelling in them as part of his or her body. When we watch the performer, we try to correlate these moves by seeking to dwell in them from outside, and we are able to dwell in the moves by interiorizing them. In order to achieve this, we need to integrate the various levels of which each one of us is composed, including our whole body. Polanyi speaks of the hierarchy of biotic levels: "The vegetative system, sustaining life at rest, leaves open the possibilities of bodily movement by means of muscular action, and the principles of muscular action leaves open their integration to innate patterns of behaviour. Such patterns leave open, once more, their shaping by intelligence, the working of which offers, in its turn, wide-ranging possibilities for the exercise of still higher principles in those of us who possess them."[23] The healthy person focuses on aspects of the surrounding space, attaining access to the world. But people afflicted with disease may become focused on the inner body itself, with the result that they lose coherence; they lose their ability to reach out beyond themselves in order to grasp the organized wholeness of the surrounding composition within which they move and act.

## SUMMARY

Why are we so estranged from nature? This is the question that has guided our inquiry into the relationship of human beings and nature. We have pursued the ideal of the knowledge of nature in the naturalistic scientific tradition. On the basis of Descartes' critical philosophy, we think of knowledge as being fully explicit. Knowledge is considered to be based on verbal conceptual thought. By distinguishing between the primary and secondary qualities of objects, John Locke reinforced Descartes' critical attitude. Locke argued that since the sensations we have of the object are not part of the object itself, they are secondary rather than primary qualities of the object, and the ideas of the object arising in us on the basis of sensations are not resemblances of the object. Sensory experience gives us merely an apparent world in contrast to the real world; the knowledge of the world must be based on abstract thinking in terms of the primary qualities, such as bulk and motion.

In this chapter, we focused increasingly on the question, How can we overcome our estrangement from nature? In this context, the critical attitude towards sensory experience, which comprises the exclusion of vision in favour of thinking, needs to be re-examined. Rudolf Arnheim tells us that any kind of creative thinking is a shaping of visual images. Thinking requires, as its object, images from the world of experience. Our thinking cannot go beyond the structures our senses supply us with. Concepts and thinking consist of the working up of images. Thinking applies concepts in the realm of perception itself. Perception and conception further one another because they are two aspects of a single unitary experience.

We have already countered the dualism of mind and body founded by Descartes by speaking of the body subject. The body subject reaches out beyond itself and forms a meeting ground with aspects of the world. Michael Polanyi has proposed in his theory of personal knowledge that we incorporate into the notion of knowledge the part that we ourselves necessarily contribute in shaping it. Knowledge is not explicit; there is also a tacit component in knowing, inasmuch as we know more than we can tell. There are constituents of knowledge that are unspecifiable. They are unspecifiable because they are personal. Knowing entails the apprehension of a whole in terms of its parts, and human beings are able to achieve this integration. We understand the joint meaning of the parts within the context of the whole only by dwelling in them. This requires that we engage our whole body because our body is the ultimate instrument of all external knowledge. Knowing is a skilful performance achieved by a person committed to wanting to know.

### CHAPTER TEN

# Nature and Culture, 2

## CULTURE IN TERMS OF THE RECIPROCAL ACTION BETWEEN HUMAN LIFE AND FORMS CREATED BY PEOPLE

In chapter 5, the first chapter on nature and culture, the discussion led us to the conclusion that the composition of spaces created by humans can severely hinder the life of natural beings, as we find them in the state of the wilderness. In that chapter, we defined culture simply in terms of what human beings have made of natural things or what they have permanently added to nature.[1] This would include the "technocentric" mode[2] of viewing the environment, of which Timothy O'Riordan speaks. It will be recalled that this mode is based on the notion that the natural environment consists of objects that need to be explored because they may be useful to the material life of human beings. This mode of viewing the environment requires an objective, rational assessment of natural resources as well as their management.

Since reason gives objects a definitive and permanent form, it is possible to assess the environment objectively. But we must not only speak about the objective assessment of the environment but must also consider the quality of the environment. Culture must not merely be defined in terms of what human beings have made out of natural things or what they have added to nature. Such a broad definition of culture includes works produced by humans which are harmful to natural things as well as to human beings. Culture needs to be understood in a more focused manner in terms of the reciprocal action between human life and objective forms created by humans. Accordingly, sociologist Georg Simmel defines culture as the path of the soul towards itself, the tool for its self-realization.[3] The soul contains within itself the preformation of its own perfected form, a form that is not yet realized within the present. The path towards its realization requires the creation of objective mental forms. The reality of these forms, though, does not merely remain a content of the human mind whose action created it. Instead, it becomes part of an independent objective entity, such as a work of art. These forms do not only take on an independent existence. People may submit to their presence and absorb their content so that it becomes effective within them and helps them to become what they have not yet become.

An idea can arise from the sphere of the intellect alone. When this happens, the idea is completely conceptual in nature. Scientific conception has as its aim the discernment of a common lawfulness underlying nature. Knowledge of nature as the factual external world is considered to be fully explicit. Explicit sentences may, however, be largely incapable of expressing what is meaningful in human terms, so laying hold of the external world by the intellect alone may have very little impact on the person who accomplishes it. But an idea can arise not only from the intellect; it may have its origin in both the intellect and the soul. In this case, it may rule our whole attitude, deeply affecting our life and what arises from it.

Here, we must familiarize ourselves briefly with Carl Gustav Jung's terminology concerning the structure of the psyche.[4] Jung held that the psyche is not equivalent to consciousness; there is also the reality of the night-world where consciousness no longer reigns. Consciousness is but one aspect of the psyche. Jung distinguished a number of "limited functional complexes" that comprise the psyche. These are the intellect, the soul, and the spirit. The intellect is clearly the power of conscious thought, but the psyche includes the soul as well as the intellect. Jung distinguished three aspects of the soul: the conscious inner attitude, the personal unconscious, and the collective unconscious. The conscious inner attitude is mainly directed towards the situation as it exists within the present. The personal unconscious contains what was formerly in the conscious but has left it, or what may never have entered the conscious, even though these items have entered the psyche. (This may be the case, for instance, with sensory impressions that lack the intensity needed to enter consciousness.) The collective unconscious contains images that are potentially present in all of humankind. Jung speaks of them initially as "primordial images" and later calls them "archetypes." They are the actual content of the collective unconscious. These archetypal motives manifest themselves in each individual life in the form of myths and pictures, such as those of the gentle and the dangerous animal.

The spirit, according to Jung, encompasses both the soul and the intellect, linking the two into one unit. He states: "Spirit, like God, denotes an object of psychic experience which cannot be proved to exist in the external world and cannot be understood rationally. This is its meaning if we use the word 'spirit' in its best sense. Once we have freed ourselves from the prejudice that we have to refer a concept either to objects of external experience or to a priori categories of reason, we can turn our attention and curiosity wholly to that strange and still unknown thing we call 'spirit.' "[5] We are shaped not primarily by concepts arising in our consciousness but by images of our life arising in the unconscious; we are dealing with mythic apprehension as well as scientific conception. Mythic apprehensions are not conceptions arising within the intellectual structure but are apprehensions arising from the spirit, and they may shape the course of our life.

In Jungian terms, the religious apprehension of the personal god who created humans in his own image in our Western religious tradition is an acquirement of the spirit. In Simmel's sense of culture, such an apprehension is a cultural form because the idea attains an independent existence and those who submit to its presence may attain a higher level of selfhood. We need to ask how a higher level of being arises from a lower level. Referring to different levels of organic life in nature and the production of more complex forms from simpler ones, Darwin spoke about "evolution." Darwin saw the world as a world of chance and evolution as the result of a struggle for existence and the survival of the fittest. For Darwin, the principles of the evolution of new forms of life were inherent in nature. Nietzsche rejected Darwin's interpretation of evolutionary development as applied to human life. According to Nietzsche, living beings have an instinctive drive to reach beyond themselves. Contemporary humankind, too, has an inherent drive to realize itself in a higher form. What value implies for Nietzsche, then, is the creation of conditions that allow this drive to become realized. Both preservation and the enhancement of that which exists as a whole must be achieved.[6]

Polanyi calls the attainment of higher levels of being from lower levels "emergence." As noted in the previous chapter, a higher level can come into existence only through a process that does not exist at the lower level. The individual item of the lower level must therefore emerge into a new coherent pattern which forms the next upper level. There exists always a gap separating a higher from a lower level of being. In the emergence of the life of the individual, the cultural forms are necessary to bridge the gap, and this is achievable only by the individual person.

## CIRCUMSTANCE

The emergence of the "being" of the individual is not possible in isolation. The individual must relate in special ways to his or her circumstance – to that which surrounds the person. The self gives meaning to its circumstance and acts in relation to it. We find ourselves, therefore, to be always in a situation. In his essay "The Morality of Music," Rudolf Kassner has contrasted knowledge, conviction, and principle on the one hand and situation on the other. He says, "Yes, I am in Rome, I am a stranger and I do not seek anything which separates me from the things. I do not seek any contract, law, or reason which may place themselves between myself and the things." Thus, reason by itself merely judges, and it does not have the things. Kassner continues: "What interests us in a foreign place is the relationship and not the contract. It is because of this that we find ourselves among the things, between the loved one and the one who loves, right among them without beginning and without end."[7]

As we saw earlier, Kassner speaks about a state of perfection that people create for themselves together with a mirror in which their dreams become alive. This state of perfection together with the mirror they may call "the golden age" or "paradise." Now, we recognize that instead of being in this state of perfection, we find ourselves in a situation, right among the things. Jung speaks of a process of anticipation or intuition, which he considers to be one of the basic functions of the soul. Through this process, we avail ourselves of the possibilities that lie within a situation.

When we learn to understand ourselves in relation to the "other," we have whatever we make present to us, our circumstance. The circumstance is what surrounds us. Ortega y Gasset interprets the relationship between ourselves and aspects of the world by stating that the object and I stand confronting one another, and although one is outside the other, the two are inseparable from each other. Ortega says in his *Meditations on Quixote*, "I am myself plus my circumstance, and if I do not save it, I cannot save myself."[8] Ortega's position is in contrast to both realism and idealism. When we consider the relationship between the subject and object, between consciousness and what we are conscious of, the world may imprint itself on consciousness; the subject has no choice in this matter. This relationship is called realism. The opposite to realism is idealism. While in realism things come to consciousness from without and imprint themselves physically and inevitably on the mind, in idealism things are contents of consciousness and come from within; they are ideas that arise in the form of thinking. What we think is the only certainty we have. It is something that exists in itself and by itself and is composed of logical matter.

In his first meditation, entitled "The Forest," Ortega shows that without me, there is no forest; the forest is not in itself, as in realism, nor is it merely a content of my consciousness, as in idealism.[9] The forest is something that is distinct from me, but it needs me in order to be the forest. But I in turn need the forest to be myself and to exist, because I am both "I" and "the forest" as one aspect of my circumstance. What does Ortega understand, then, by "circumstance"? He writes: "Circumstance! *circum stantia*! That is the mute things which are all around us. Very close to us they raise their silent faces with an expression of humility and eagerness as if they needed our acceptance of their offering and at the same time were ashamed of the apparent simplicity of their gift. We walk blindly among them, our gaze fixed on remote enterprises, embarked upon the conquest of distant schematic cities."[10] Circumstance is what I make, but I make it with the things because I have built a relationship. When Ortega speaks of "the mute things which are all around us," he refers to our circumstance which has no voice, yet these same things with "their silent faces" need us, and they want to draw our attention to them. But instead of directing our attention to them – entering a relationship with them – our eyes are fixed on remote enterprises.

### BRINGING INTO PRESENCE

Nature is that which brings itself forth out of itself, and culture is what we as human beings have made out of nature or what we have added to it. We need to consider also in our understanding of culture the impact that the works of humans may have on the quality of human life. Martin Buber has inquired into the principles of human life in general. He compares human life with other forms of life and concludes that human beings emerge as a result of two kinds of movement. The first is a movement that sets us at a distance from the things. Once this distance has been attained, the second movement can commence, and in this way we enter into a relationship with the things.[11]

In order to enter into a relationship, the entity we relate to has to attain an autonomous state and what we relate to has to be present. We have to bring it into our presence. How we bring something into our presence can be illustrated by means of the discovery of landscape art. We take the experience of landscape and landscape art for granted, but Rilke tells us that landscape art took a long time to emerge: "In the pictures on vases ... the surroundings are only mentioned, abbreviated and given merely in the form of the initial letters ... Although having already existed for thousands of years, man was still too new to himself, too delighted with himself and hence not yet able to see beyond and away from himself."[12] After describing Greek art, as expressed in this form, Rilke speaks about Christian art: "[In Christian art] humans were [like] tunics and became bodies only in hell, and the earth was only rarely allowed to be the landscape. Wherever it was lovely it had to be heaven; wherever it caused fright and became wild and inhospitable it turned into the place of those banned and lost forever."[13] We are reminded how Petrarch, after climbing Mont Ventoux and reaching its peak, discovered the space of landscape and at that very moment became frightened that he might lose himself. The profound experience of the landscape was not permitted to occur; as Rilke points out, heaven and hell were the only acknowledged realities.

Then Rilke tells us about Leonardo da Vinci's landscape art: "When one painted the landscape one did not mean *it* but one's own self. Landscape had become a pretext for human feeling, a parable of man's joy, simplicity and devoutness. It had become art. Leonardo had already adopted it in that form. The landscapes in his pictures are expressions of his deepest experience and knowledge, blue mirrors, in which secret laws reflect upon one another, distances as vast and undeciphered as the future."[14] Through the physiognomy of the Mona Lisa within the landscape composition, Leonardo described the profound experience of landscape.

Rilke goes on to describe how landscape finally emerged as an entity within itself which we learned to experience as such: "Art in form of landscape developed in the direction which Leonardo da Vinci had already intuitively possessed. It formed itself slowly in the hands of lonely men throughout hundreds of years. The path which had to be taken was long and arduous since it was difficult to so distance oneself from the world that one no longer saw it with the prejudiced eye of the citizen

Fig. 4    Meindert Hobbema, *The Road to Middelharnis*, 1689 (National Gallery, London)

born here who sees everything in relation to himself and his needs."[15] According to Buber, human beings emerge by the movement that distances them from things and by the movement that brings them into a relationship with things. Bringing landscape into our presence requires both of these movements.

As we have seen, in naturalism it is our aim to imitate the outer appearance of the objects pictorially. In order to achieve this, we take a central position and the objects are opposite us and apart from us. In terms of the pictorial representation of the relations between space and body, we have been taught to view and represent visual space, such as the space of the landscape, in central perspective projection. A good example is the painting *The Road to Middelharnis* by Meindert Hobbema (figure 4). In this painting, straight-line relationships govern the construction of the pictorial space of the landscape. In using the central perspective projection, as Hobbema has done here, one aims to render three-dimensional objective space and objects as they are placed within that space on the plane surface of the canvas. Referring to the central perspective projection, the artist Jesus Rafael Soto stated: "An art critic once said to me that it was understandable that people in the Renaissance should have marvelled at a painting of a tree in a particular space; the space was the third dimension. It was like a window. A gentleman could sit as in a theatre picturing whole situations which he was no part of. The killing or kissing or lovemaking went on in one place and he stayed in another."[16]

Returning to the movements that Buber identifies as necessary to reach the principle of human life, one has indeed attained distance from the things. A particular space is three-dimensional and part of one objective space. Accordingly, the space forms a homogeneous continuum and the figure in the space – for instance, the tree – refers to that continuum. Naturalism considers art as objective cognition – it is based on scientific fact, on the positivist's outlook on the unchanging nature of the object. Rendering in the form of a photograph is thought to be an objective statement in virtual space because it is free of the prejudices of a person. A photographic rendering of an aspect of the environment implies that the viewer takes a fixed position and that what is represented is bodies in space. It is a rendering of aspects of the objective world in terms of the stationary visual sense.

Central perspective projection is not the only mode in which we can experience aspects of the world. We experience the movement of the points of light in darkness, which I discussed earlier, as if it were taking place in a space that is concave and shallow, a spheroidal space. That the

Fig. 5    William Turner, *Rain, Steam and Speed*, 1843 (National Gallery, London)

structure of visual space can be experienced in that form, which is so very different from the experience of space represented in perspective projection, is born out in certain paintings, such as some of William Turner's. In contrast to the straight-line relationships in Hobbema's landscape, Turner's *Rain, Steam and Speed* (figure 5) is vortexlike, dominated by curved lines. What we perceive is evidently not something pregiven in itself but something that we must actively integrate.

We do not live in the objective world, however. We live in a life-world, and we participate in it by seeing and moving. Our self-movement causes our surrounding world to move too, and it moves in curved space, as do the moving points of light in darkness when our own position is fixed. The straight-line relations of the central perspective projection give way to curved space. In the space of Turner's picture, we become engaged in the world rather than remaining mere spectators.

The relationship between the artist and aspects of the world that prevail in naturalism changed profoundly around the turn of the twentieth century. For the painter Franz Marc, for instance, the outer world of appearance is merely a superficial aspect. His predominant motif is the animal in the landscape. He strives to let the inner reality of the beings, in this case the animals, reveal itself to him. In opposition to naturalism, he calls his own means of achieving this goal "animalization." If naturalization, a process based on naturalism, means the imitation from the outside, as seen from the standpoint of the spectator looking at a spectacle, animalization means dwelling in the animal, elucidating the animal-like nature of the life-space of the animal.[17] Marc recognizes the animal's life-space as distinct from his own. He describes his aim when he portrays the animal within its own surrounding: "We will no longer paint the forest or the horse as they may please us but as they truly are, as the forest or the horse feel themselves, their essence behind appearance, the only thing we are able to see."[18]

While in naturalization we form the centre, as in the central perspective projection, in animalization we must abandon the position of centre. We must extend ourselves to encompass other things and beings. The dethronement of the self through abandonment of the central position permits Franz Marc to grasp the significance of the animal within the landscape. It is not merely a body in space; instead, it takes its own place. The naturalistic painter focused from his fixed position upon the figure, such as the animal, and often lost sight of the relationship of the figure to the ground – the animal in the landscape. We can say, then, that more basic than the mere perception of form is the distinction between figure

Fig. 6   Georges Braque, *Maisons à L'Estaque*, 1908 (Hermann-und-Margrit-Rupf-Stiftung, Musée des Beaux-Arts de Berne, Switzerland)

and ground. While we used to view the ground as empty, we now learn to recognize the ground as the medium that harbours. Furthermore, the artist's position is no longer fixed, nor is the position of the source of light, and this profoundly affects the figure-to-ground relationship. The figure affects the ground, as does the ground the figure, and hence figure and ground belong together.

The construction of novel figure-to-ground relations, and hence new experiences of space, is made possible through the techniques invented by the painters of cubism. In Georges Braque's *Maisons à L'Estaque* (figure 6), houses that would recede from foreground to background in perspective representations of landscape are instead all piled up in the plane of the picture in the form of simple cubes;[19] the landscape appears to be illuminated from the interior of the houses themselves, and the colour tones are distributed more or less evenly over the whole painting. As a result, the figure-to-ground relationship becomes obscure, giving rise to an experience of space, which we ourselves have entered. In the construction of cubism, instead of space as such that contains objects, there are aspects of objects that create space.

A picture is not merely something that deals with what is concrete in terms of physical reality. According to Martin Buber, art is the work and witness of the relationship between the *substantia humana* and the *substantia rerum*; it is the realm of "the between" that has become form.[20] The painting of a landscape is a realm in which the emergence of the relationship between humans and things takes on form.

### SUMMARY

This chapter continued the discussion about the relationship of nature and culture. Culture is not merely what we have made of natural things or what we have permanently added to nature. As Georg Simmel has pointed out, culture consists of those forms created by human beings that assist humans in becoming what they are meant to be but what they have not yet become.

In Jungian terminology, both the intellect and the soul (the entity called "spirit") must originate the idea that underlies a cultural work, such as a work of art. We recognize different levels of "being." The principle that rules a higher level of being is not present at a lower level. Thus, there exists a gap between two different levels. When we speak of the emergent principle, we therefore imply the closing of this gap, which is achievable only by the individual person.

We create for ourselves a state of perfection together with a mirror in which our dreams come alive. We might call such a state "the golden age." From our contemporary perspective, this state of perfection seems to have vanished. We now find ourselves in a situation, right among the things, and it is therefore extremely important for us to find out how these things are. In order to recognize the "being" of these things, as well as our potential relationship with them, we have to be able to bring them into our presence. How one brings something into one's presence is illustrated by means of the discovery of landscape art, as one cultural form. It requires that we attain distance from these things and then enter into a relationship with them. For Martin Buber, art is the expression of people's relationship to the things; it is the realm of "the between" which has become form.

### CHAPTER ELEVEN

# Nature as Landscape

## WHAT IS LANDSCAPE?

The knowledge of nature is not limited to functional relations, which have come to play such an important role in the scientific study of nature, but it extends also to spatial relations. These spatial relations may in turn be very different, depending on whether they are thought of in terms of the objective world or whether they are experienced in everyday life. This can be illustrated by considering, once again, landscape as an aspect of nature. The English word "landscape" is an old Germanic term. Grimm's German dictionary refers to the Old High German form "landscaf" and gives various definitions, the most common being: "1. Tract of country, land complex with regard to location and natural condition, in new sources especially with regard to the impression which such a tract of land makes upon the eye, and figuratively: the flowering landscape of our life; and 2. for that reason and already found in old sources, the artistic pictorial presentation of such a tract of land." [1] A distinction is made between landscape as the actual land as it is in its own composition, regardless of whether we as experiencing persons are present or not, and landscape as the experiential space of everyday life which requires the presence not only of the actual land but also of ourself with our particular point of view. The dictionary's concluding definition of the word states that landscape exists in the form of images or their representation in the form of pictorial art.

When we seek to obtain the knowledge of the one objective world encompassing all nature, we usually recognize but one objective space and one objective space-time framework. We can therefore construct in symbolic form the geometry and composition of spaces much larger than those that we take in at a glance from one point of view. This is done, for example, in J. Brian Bird's *Natural Landscapes of Canada: A Study in Regional Earth Science.*[2] The one and only time scale is given here in the form of the continuous scale of geological history. The one and only scale of space is implied as follows: "In the present work, a three-fold division [of Canada] into the Canadian shield, the border plains, and the extensive mountain belts, with a fourth small Arctic coastal plain, is considered more appropriate for descriptive purposes."[3]

We can dissect in thought the spatial composition of a landscape in terms of its elements and can determine which elements are present and how they are situated in surface features of the terrain. We can also determine characteristics of the vegetation and how the vegetation is structured. We may wish to explain the natural composition of the objective landscape in terms of the natural forces that are effective in it, and to elucidate on this basis the reasons for the particular composition of that landscape. What scale, though, do we have to consider in order to describe and analyse a natural landscape? Reference to the earth as a whole appears to be desirable, as explained below.

Natural landscapes are structured and determinate in their composition. Their structure and their composition in terms of plant and animal life-forms are determined by their geographical location on the earth. Landscapes situated on the same latitudes in the northern and southern hemispheres can be very different, for there are great differences between the two hemispheres in the amounts of land and sea. In the southern hemisphere, vast areas are permanently covered with water and an oceanic climate prevails. By contrast, there are continental climatic conditions in the northern hemisphere. Differences in the amount of land and its distribution give rise to great differences in the temperature regime on land. In the north we have a seasonal climate, where the greatest differences in temperature during the year at any particular location are expected between seasons, such as summer and winter. In the southern hemisphere, and also at higher altitudes in the tropics, the greatest differences in temperature throughout the year are recorded between day and night. This is known as a diurnal climate.[4]

Such climatic regimes of course have an impact on the form of vegetation present in a region. Think of a savanna and a tropical rain forest and their prevailing climatic conditions and how very different the life-spaces are in these two different landscapes. Would one not also expect to find very different life-forms of animals? The forms characteristic for a savanna would of course have different surroundings, or realms, from those for a tropical rain forest. Similarly, the whole invisible network of interconnections of meaning, as they are shaped by the presence of the animals, would be characteristic for each region and would be different in a savanna from those in a tropical rain forest. The structure of the earth, especially in terms of its surface features, determines the natural composition of the landscape, including the nature of the surroundings of animals through their respective animal life-form. It is a determinate structure and we, as human beings, are not present.

But does this situation change when we are present? We need to consider here the nature of our human experience of landscape. The question arises whether there is a direct relationship between the objective structure of landscape and the nature of our experience of it. The geographer Jay Appleton has attempted to deal with the "experience of landscape" – how and why we get pleasure from the perception of certain kinds of landscape. Why do we like one landscape and dislike another? And, correspondingly, why do we act in one manner in relation to one landscape and differently in relation to another? Appleton has explained our aesthetic sense of space of the landscape with a "habitat theory."[5] He suggests that there are two aspects basic to our appreciation of a landscape; he calls them "refuge" and "prospect." It is the desire, intrinsic in all of us, that at least at times we wish not to be seen: we like structural elements within a landscape that allow us to hide. This is the sensation of refuge or its representation in symbolic form, such as in a landscape painting. We like to experience, from a safe distance, the majestic powers within nature – hazards that could destroy us. We wish also to see far ahead and we therefore enjoy the distant horizon – the prospect. When aspects of our surrounding world are hidden by various structures, we are able to move about within that landscape, exploring what lies behind those structures, beyond them.

Why, then, do we like a particular landscape? We do so because it affords us the pleasure of fulfilling these innate desires, not necessarily all at once but at appropriate intervals within our encounter of that landscape; because it furthers our basic biological needs of maintaining ourselves as well as our species. The pleasure various landscape types give us is thus the result of the affordances they offer for fulfilling our basic needs – to be able to see and not be seen, to be able to look beyond, and to sense nature's hazards at safe distances. Appleton intends his habitat theory to account for why we like one landscape and dislike another, to discover the causes in order to shed light on the aesthetic experience of landscape in general. He explains the aesthetic preference of one landscape over another in Darwinian terms – in terms of the affordances the landscape offers us for our survival. We should note that much of Appleton's pictorial evidence consists of "picturesque" landscape paintings. As well, he chooses perspective photographs of contemporary landscapes, of houses for hiding and paths designed to make us continue walking in order to explore what lies hidden in the near distance. They are landscapes as viewed when we form the centre, as in the central perspective projection.

Appleton thinks about the composition of landscape in geographical terms and about the impact the geographical features of the landscape have on our survival. He refers to this as "experience." According to his habitat theory, the geographical features of a landscape determine the nature of our experience of the landscape. Let us inquire into this supposedly close relationship between objective structure and experience by examining the experience. The psychiatrist Fred Fischer speaks about the "path experience," that is, the experience of our surrounding space while we as the subject are moving directionally along a path.[6] A path experience differs greatly, depending on whether we are moving along the path from home towards a distant goal or whether we are on the same path on our way home after we have attained the goal or after we have abandoned it. Even though the landscape is geographically the same, our experience of it is entirely different.

How different the experience of a landscape may be, depending on our attitude, is illustrated by J.J. Winckelmann. When crossing the snow-covered Tyrolean Alps on a trip to Italy in 1755, he was deeply impressed by the magnificence of the landscape, and he determined to return one day in order to enjoy once more this delightful experience. After remaining in Italy for thirteen years, immersing himself in the study of antiquity while surrounded by the urbane Italian culture, Winckelmann returned by way of the Alps in 1768. This time, the sight of the mountains, their excessive height, and even such human creations as the farmhouses in these mountainous regions, horrified him.

The experience of landscape and its valuation are not based solely on biological expediences, as proposed by Appleton, but upon our whole multifunctional nature, including aesthetic pleasure and the pleasure of understanding. We cannot reduce one function to another. The experience of the mountain, for example, may be that of the sacred mountain – the place of the gods – and the path to the mountain may be the experience of the pilgrimage of the path subject.

## NATURE AS LANDSCAPE

Nature in our absence is a determinate structure. From the scientific standpoint, too, nature is regarded as the objective realm of the external world. In our inquiry into nature within this context, we accept the role of spectator or impartial watcher. We are the outsider and we regard human beings, too, as part of the bodily world that is independent of the experiencing subject. The knowledge of nature that we acquire in the context of science is independent of the person who accomplishes it.

But in our everyday experience, in our life-world, we are present together with aspects of nature. In our everyday experience we shape aspects of nature into landscape. While the objective study of nature discards this experience as inconsequential, "nature as landscape" considers it to be important. "Nature as landscape" adopts the view that any understanding of our relationship to nature must be founded on perceptual experience. In "nature as landscape" a world emerges which, in contrast to "nature," is not a determinate world.

It is part of our human condition that we have to create a world. This is described for us so well in Jung's *Memories, Dreams, Reflections*:

From Nairobi we used a small Ford to visit the Athi Plains, a great game preserve. From the low hill in this broad savanna a magnificent prospect opened out to us. To the very brink of the horizon we saw gigantic herds of animals: gazelle, antelope, gnu, zebra, warthog, and so on. Grazing, heads nodding, the herds moved forward like slow rivers. There was scarcely any sound save the melancholy cry of a bird of prey. This was the stillness of the eternal beginning, the world as it had always been, in the state of non-being; for until then no one had been present to know that it was this world. I walked away from my companions until I had put them out of sight, and savoured the feeling of being entirely alone. There I was now, the first human being to recognize that this was the world, but who did not know that in this moment he had first created it.[7]

"Being," such as the "being" of a landscape, is something that you and I make, but we make it together with the things. Discovering the "being" of beings requires that we bring them into our presence by finding our place among them. Jung had to leave his companions in order to be by himself but together with the things composing the landscape. When he brought the landscape into his presence, this aspect of the earth was no longer determinate, as it had been in his absence. The world that emerged in his presence was a possibility and not a necessity, similar to the reality of the game of chess, described earlier, which the two players created.

In his habitat theory, Appleton proposes that our experience of landscape is necessarily determined by the structure of the landscape. In *Place and Placelessness*, Relph reminds us that experience is intentional, and that character and meaning are inscribed into the landscape by the intentionality of experience.[8] Our experience of landscape thus arises both from the structure of the landscape and from our intentions in relation to it. In "nature as landscape" we cannot maintain the dichotomy between the objective world and the subjective world. Such a dichotomy implies that we are outsiders who do not participate in the world. But the world we create is the outcome of the relations between humans and the things. August Nitschke speaks of "configurations," which he describes as "the manner in which figures – persons, animals, plants, stones – are positioned in relation to one another in a picture."[9]

Configurations exist both in the imagination of persons and in the world they create. Wolf Jobst Siedler has aptly expressed this linkage: "Long before the Romantic Rebellion had turned laws in literature upside down, it had transformed gardens and parks into an elaborate wilderness. Yet another ten years, and Marie Antoinette was drawn toward pond and meadow, the Mill of Hameau replaced Grand and Petit Trianon. Then the Revolution set in."[10] Siedler refers to the triumph of the English garden over the French garden style as a mark of a correspondingly profound change in people's imagination. The geometrical design of the French garden was seen as a reflection of human beings, who in their original state had been innocent and virtuous but who were now corrupted by culture. People needed to re-enter nature; the image of nature as it brings itself forth, penetrating the formal garden, began to possess them.

Configurations are based on the functional forces arising within the individual (or the group of which the individual is a member) and the structure of that person's life-space. As a rule, several functions are simultaneously present in an act. Functions must not be projected into an object but must be considered foremost with the individual person as their source. When we consider the aesthetic function, for example, we are concerned with the aesthetic as a source of human activity and its creations.[11]

## LANDSCAPE AS PLACE

As noted above, "nature as landscape" concerns itself with a meeting between human beings and aspects of nature, giving rise to a world that is a possibility and not a necessity. A landscape is not merely an objective structure; it also makes an impression on us and has an emotional impact on us. It is, in fact, our sense of feeling that makes us open to the world. When I am glad, I more readily gain access to the "being" of the landscape in which I find myself.[12]

It will be remembered that Jung said that the psyche is composed of the soul as well as the intellect, and that the integration of the soul and the intellect represents the spirit. The collective unconscious, as one aspect of the soul, is shaped in the form of archetypes. The anima is the most important archetypal figure and is connected both to the collective unconscious and to the personal psyche. Jung explained the anima as follows: "I call the outer attitude, the outward face, the *persona*; the inner attitude, the inward face, I call the *anima* ... As to the character of the anima, my experience confirms the rule that it is, by and large, *complementary* to the character of the persona. The anima usually contains all those common human qualities which the conscious attitude lacks."[13]

Through this figure, the soul links us to the place on earth where we live and lets us act in relation to its composition. The soul can therefore be understood only in terms of our being dwellers who inhabit the earth, for the earth has shaped the soul. There is, for example, the image of "mother" – that which helps us live our own life. Just as the mother helps us, so does the image of "earth"; for, as her inhabitants, we sense our profound dependence on her. What the mother has been for the child, the figure of "mother earth" may become for the adult, but only our spirit can make us experience our relationship to the earth in this manner.

Neil Evernden, in an article entitled "Beauty and Nothingness," describes the passengers on a train crossing Canada as it passes from the Precambrian shield onto the central plain. It is nearing dusk and very few travellers have emerged from their club cars and roomettes to go up to the observation dome. The conductor is not surprised, because for him "there's nothing to see."[14] Those who do gather in the observation dome are from the prairies; only they appear to find the prairie compelling.

Images of place corresponding to the prairie landscape are present in these passengers. But the travellers not raised on the prairies lack these meaningful images, and their objective standpoint sees only an emptiness holding nothing, a landscape destitute of any visual merits. According to objective criteria for the valuation of different kinds of landscapes in terms of their visual merits, the prairie has no visual merits. What evidently characterizes the prairie is the absence of things. But Evernden tells us that "the prairie is an experience, not an object."[15] For the passengers who come from the prairies, the prairie is a dwelling place. We may say that, for them, the prairie is emptiness which holds, which harbours.

The prairie is a natural landscape that has shaped itself and is permanent in relation to the span of human life. Superimposed on the prairie as this natural and self-building form was the settler, who had to build the settlement, a form shaped by people. The built settlement as cultural form is the point of arrival, and it acts as a centre, generating fields of forces that express themselves in the form of paths. These directional paths subdivide the seemingly infinite extensions. The ordered space that emerges admits particular patterns and rhythms of action. In order to build places, the settlers had to recognize the character of the natural space.

We are reminded once more of Simmel's understanding of culture as the reciprocal action between forms created by humans and human life, the tool for human self-realization.[16] As inhabitants of the earth, our souls are shaped by the earth, but the emergence of the soul towards what it is meant to be requires the creation of cultural forms, such as the settlement, as they complement the structure of the soul. This was indeed the case for the train passengers who had come from the prairies, as image and sense of feeling were joined. Dwelling implies the presence of place, and as the architect Norberg-Schulz has shown, place consists in orientation and identification.[17] Both ordered space and built form make it possible for us to know where we are and to gain a sense of belonging to a place.

## SUMMARY

We distinguish between landscape as an objective spatial structure and landscape as the experiential space of everyday life. Objective landscape is determinate in its composition. According to Appleton's habitat theory, our presence in the landscape does not alter the determinate characteristics of the landscape. Appleton also holds that our experience of the landscape is determined solely by the structure of the landscape, through the affordances it offers us for our survival. Winckelmann's reaction to the mountainous Tyrolean landscape on his return trip and the path experience described by Fischer demonstrate that we may experience the same landscape differently, depending on our intentions. Thus, our experience of landscape may be determined both by the geographical structure of a landscape and also by our intentions.

Nature in our absence is a determinate structure and it remains determinate when we are merely observers, or outsiders. But when we enter nature by shaping it in the form of "nature as landscape," it becomes an indeterminate world. A world arises that is a configuration, and a configuration is a possibility and not a necessity. Finally, through the archetypal figure of the anima, our soul is bound to the earth. Our image of place is shaped through the nature of the space with which we first become familiar.

CHAPTER TWELVE

# Construction of a World

## TWO DIFFERENT WORLDS

While the animal's surrounding realm is fixed through the schemata present in its inner structure, humans have gained a symbolic imagination that enables them to build different worlds.[1] We are born into a life-world that is composed of aspects of nature, other human beings, and various works of human beings. As we have seen, what surrounds us shapes our preference as well as our sense of home, out of which we build a world. In *Place and Placelessness*, Relph stresses the importance of the relationship between the experiencing person and the experienced place: places are significant centres of our immediate experience of aspects of the world.[2] Concerning the significance of our experience of landscape for the constitution of our world, Ortega y Gasset writes: "My natural exit toward the universe is through the mountain passes of the Guadarrama or the plain of Ontigola. This sector of circumstantial reality forms the other half of my person, only through it can I integrate myself and be fully myself."[3] The mountain passes of the Guadarrama or the plain of Ontigola thus form part of Ortega's circumstance.

Experience of place belongs to our life-world, and the life-world is always pregiven. While "environment" has come to refer to the objective world, "circumstance" is an integral part of our life-world. When we bring something into presence and we learn to understand ourselves in relation to what is present, we have "circumstance."[4] It is evidently through the experience of particular landscapes that we have access to the world, insofar as these landscapes form part of our circumstance. There are at least two very different modes of obtaining a world. The expressionist writer Gottfried Benn describes these alternatives as modes: "of adjoining the things to man and his natural space or of assigning the concepts into mathematical series without any contradictions."[5]

In science, the physical object is considered to be the outcome of material forces that are entirely independent of the scientist. It is the scientist's task to elucidate these forces in an objective sense. In order to attain the knowledge of the objective world and the laws underlying it, we have to invent mathematical space and construct the modern physical theory of nature. Mathematical objective space is something that is by itself. In objective space, we treat locations as mere positions and we treat the space between locations as measurable distance – as *spatium*, or interval.[6] The locations of things have no particular significance but are interchangeable.

In art, the manner of viewing, including our own position, is an important quality in our discovery of the "being" of what *is*. When an artist such as Cézanne has decided on a particular landscape motif, he must restructure the whole visual content in such a way as to lead to an order that he has discovered and clarified. A picture is not merely a dealing with what is concrete in terms of physical reality. The painting of a landscape is a realm in which the emergence of the relations between people and the things take on form.[7] Cézanne believed that he merely reproduced nature in his landscape pictures. He chose first a motif – a particular aspect of his surrounding world. He needed to apprehend his motif visually in order to find its significance within its own context.[8] In his paintings of Mont Sainte-Victoire, Cézanne realized the "being" of the mountain. In the work of art, the representation is itself a statement concerning the recognition of structure.

The painter reveals to us, the viewers, the "being" of the mountain. It is through the self-contained work of art that aspects of the earth come into our presence. We, as viewers, learn to be no longer the outsider who designates and speaks about an external world factually and explicitly, and who imitates it as the naturalist does. Through the work of art, we become insiders: by being dwellers on the earth, we learn to inhabit the world by bringing it into our presence.[9]

Instead of analysing "environment" in terms of scientific objectivity, Relph seeks an understanding of "environment" through an elucidation of the nature of our experience of place. What sets place apart from space is the experience of being inside. By means of building locations, we learn to enter the inside. Relph points out that locations can be described in terms of internal characteristics, what he calls "site," and external connectivity to other locations, what he calls "situation"; places therefore have spatial extension and an inside and outside.[10] Making space by building locations contrasts with our construction of mathematical space, which does not admit any location.

## THE CONSTRUCTION OF WORLD

The earth is the ground where we are born, but we construct the world into which we grow. The whole act of constructing comes to mind here, an enterprise about which the French writer Paul Valéry says:

Man, however great his knowledge, will never know the riches or the broad intellectual domains that are illuminated by the conscious act of *constructing* ... Constructing takes place between a project or a particular vision and the materials one has chosen. For one order of things, which is initial, we substitute another order, whatever may be the objects arranged: stones, colors, words, concepts, men, etc. Their specific nature does not change the general condition of that sort of music in which, at this point, the chosen material serves only as the timbre, if we pursue the metaphor ...

The word construction – which I purposely employed so as to indicate more forcibly the problem of human intervention in natural things and so as to direct the mind of the reader toward the logic of the subject by a material suggestion – now resumes its more limited meaning ...

Thus we can find ourselves surrounded by and moving among a multitude of structures. Let us note, for example, in how many different fashions the space around us is occupied – in other words, is formed, is conceivable – and let us try to grasp the conditions implied in our being able to perceive the variety of things with their particular qualities: a fabric, a mineral, a liquid, a cloud of smoke.[11]

In the act of construction, two very different processes may be at work. The designer Christopher Alexander refers to them as the unselfconscious process and the selfconscious process.[12] With regard to the unselfconscious process, the Expressionist painter Emil Nolde greatly admired the unity of the actions of primitive people: "Why are we artists so fond of seeing the primitive expression? In our time, for every earthen vessel or piece of jewellery, for any utensil or article of clothing, a draft on paper must first be developed. The works of primitive people grow out of the material in their hands, between their fingers. The will expressing itself is the desire and love of creating. The absolute originality, the intense and often grotesque expression of strength and life in the simplest possible form – that may be what gives us pleasure in these indigenous works."[13]

In these comments Nolde points out that in the unselfconscious process there is no mediating representation. These works of the primitives are apparently expressions of the archetypal images of the unconscious; they are images of primordial experiences given shape with the material at hand. What was hidden within the earth is brought into unconcealment in the works of these people. Alexander finds that in contrast to the unselfconscious process, the selfconscious process of construction involves an interaction between the conception of the context and the ideas and the drawings that represent the form. He recommends that we proceed beyond the pictorial representation of the form which we think may satisfy the context's demands within the ensemble. He proposes a mathematical representation of the requirements, in terms of sets, as in set theory.

How can this be applied to design, such as the design of a landscape? We wish to fit a form to a given context within the ensemble, and we want to avoid misfits between the form and the context. Misfits can be expressed as misfit variables. These variables form a set. There are various requirements in terms of the individual conditions that must be met at the form-context boundary. Within the design, such as the design of a landscape, the set may contain any requirement that serves to prevent misfit – for instance, the shape of a tree in relation to the prevailing light conditions in a particular space within the whole landscape. A set is composed of various subsets, and we must evaluate the significance of the subsets for the set as a whole. Involved, then, in the design process are both an analytical phase and a synthetic phase. Alexander refers to the latter as "the realization of the program."[14]

Regarding the logic of relations in the design and construction of structures, we may distinguish with Charles S. Peirce three categories: "Firstness is the mode of being that which is such as it is, positively and without reference to anything else. Secondness is the mode of being that which is such as it is, with regard to a second but regardless of any third. Thirdness is the mode of being of that which is as it is, in bringing a second and third in relation to each other."[15] Firstness is spontaneity, is feeling; secondness involves otherness, that which maintains a place for itself; and thirdness involves representation, continuity, and lawfulness. We can illustrate this by the development of the reality of the composition in a painting's pictorial space, which the Swiss painter Otto Meyer-Amden has expressed as "from motif/to motif form/to form motif/to form."[16]

Meyer-Amden speaks of the "extraoptical," the "involuntary inner movement," as the motif of his pictures. This is the spontaneous sense of feeling before the actual figures arise before the mind; Peirce refers to it as the category of firstness. Next, the concrete images, the actual figures or elements composing the picture come into being; they are the motif forms. This aspect of what appears before the mind but represents otherness Peirce calls secondness. In the form motif the figures are then placed together, relating one figure to the other. Form, however, comes into being only when a total congruence of the motif form and the form motif has been achieved – when their essential connection has been found. These necessary relations Peirce calls thirdness. Meyer-Amden contrasts the "involuntary inner movement" with the optical abstract composition, the "canon" that is the result of "intellect and measure." The essential connection between the motif form and the form motif is recognized, therefore, only when the involuntary inner movement is present together with the canon. Accordingly, the picture comes forth while it is located between these two fields of forces that arise within the creative person – the "involuntary inner movement" of the soul on the one hand and "intellect and measure" on the other.

## PARTICIPATION IN THE WORLD

In speaking of the objective world, we acknowledge that we are the impartial observer and that we are outsiders. While we are outsiders, the state of being inside is largely eradicated. If any inside whatsoever remains, it is separated from the outside by a threshold and it has become extraordinarily difficult for us to pass the threshold: "Pain has turned the threshold to stone."[17] A natural structure is put together solely by the forces of nature, and it is very difficult for us to enter it. In contrast to a natural structure, a place allows us to enter and to construct a world. How are we to understand this? We must inquire into the relationship between bodily things and space. When we consider a bodily thing as a mass volume, we are dealing with a closed structure. Any solid structure exists in space and occupies space, and it shapes the space around it. But there are also things or figures that are shaped out of body-forms and space-forms, out of masses and spaces. Space and body may thus inter-penetrate one another.[18]

We experience masses from outside, space from inside. Our experience of solid structures is therefore quite different from our experience of space-forms. Earlier, I referred to "pathic" and "gnostic" modes of experience.[19] The gnostic moment, or the "limited functional complex" of the intellect, tends to dominate our experience when we look at items and see them as coloured objects; whereas, in our experience of spaces, including the space of music, the pathic mode, or the limited functional complex of the soul, dominates.

In order to understand the significance of the relations between body-forms and space-forms for our experience of landscape, we should consider Jesus Soto's abandonment of the central perspective projection that lets us represent pictorial space in a way that separates the spectator from the spectacle. Soto contrasts the naturalistic tradition with his own position: "People accepted this until recently and now in my own case for example, I propose that man become immersed in a spectacle."[20] Soto is referring to his hanging sculptures, or "penetrables." The entire sculpture consists of large numbers of hanging rods arranged on a large scale into a composition in space. Soto makes the important observation that a structure consists of elements and the values between the elements – the relations. The relations are autonomous entities, which the elements (in Soto's penetrables, the rods) only serve to reveal. Soto says that he himself often speaks about "relations between elements."[21] However, the relations exist before the elements mark them out. Soto's penetrables allow people to enter them and move around in them. He has noticed that young people, especially, move around in them joyfully as if in a new medium that gives them a sense of liberation.

Soto's penetrables allow people to cross the threshold separating the inside from the outside and to enter the inside. On a much broader scale, Heidegger's world is the intercourse of what he calls the "fourfold": the earth and the sky, the mortals and the divinities.[22] These four interplay, and world arises from the interplay. The important point for us is that humans, as mortals, are not outside of "being" but form a part of it; we are in the world (*In-der-Welt-sein*).[23]

Like the structures of Jesus Soto, in which the elements reveal the relations, a landscape is composed of elements that bring the relations into openness. When we design a landscape, we design the emptiness that holds,[24] the relations, and the elements composing the landscape serve to reveal the relations. Just as we are able to enter one of Soto's structures, so we can enter a landscape. When we discover the appropriate relations, the landscape may become a dwelling place for us. To be able to realize a dwelling place, we must not only recognize the quality of an existing composition, but we must also be able to foresee the quality of a spatial composition (such as a landscape) that is not yet realized. This foresight requires our imaginative anticipation of the future experience of living within such spaces. Among other functional forces arising within us, it is our aesthetic sense that serves in mediating while it guides our conduct or lets us anticipate our acts.

The landscape forms a part of our life-world, but we also form a significant part of the landscape. Just as the tree is embedded in the forest, so are we embedded in the landscape. To illustrate this, let us look at Franz Marc's experience. In extraordinarily difficult circumstances and under the impact of fierce fighting early in the First World War, Franz Marc wrote: "Certainly the war does not make a naturalist out of me – on the contrary: I feel so strongly the spirit which hovers behind the battles, behind each bullet, that the naturalistic, materialistic disappears altogether."[25] Marc was not an impartial watcher who could step outside the battle in order to determine its probable outcome by analysing its elements through objectifying them. He remained part of the whole battle and concentrated his imagination on the total relationship and its essential features. It was precisely the significance of the relationship within the whole setting that he attempted to grasp *while he remained part of it.* While the impartial watcher must analyse in terms of material power, Marc devoted his attention to the face of the battle, which was held together as a total event by the spirit, and in which the elements when focused on in isolation remained purely accidental.

We are reminded of the place-ballet which, Seamon tells us, occurs twice weekly in the Varberg market on the Swedish coast. When people with similar intentions come together in appropriate numbers in a supportive environment, they begin to enter this type of ballet, cohering with aspects of their surrounding space. If we wish to create together such a participatory joint action, we need to incorporate ourselves into this world and not stand apart from it.

## SUMMARY

While the surrounding world of an animal is fixed, humans through their symbolic imagination are capable of building different worlds.

With Gottfried Benn we consider two worlds: the world constructed by science and the world of the work of art as a part of our life-world. Science is composed of the knowledge of nature as the objective bodily world that is external to us. This world *is* whether we are present or not.

In the work of art, we bring nature into our presence, as we do in our experience of landscape. In order to enter the landscape, we must construct places that allow us to orient ourselves in the world. In the process of human construction, Alexander distinguishes between an unselfconscious process and a selfconscious process. The selfconscious process of construction involves an interaction between the conception of the context and the form that may satisfy the context's demands within the ensemble and avoid misfits between the form and the context. Peirce distinguishes between three categories of relations; the highest category is concerned with the essential connections.

We consider figures that are shaped out of body-forms and space-forms; bodies and space may thus interpenetrate one another. We, in turn, may now experience a solid mass from the outside or space from the inside. A landscape, too, is composed of body-forms and space-forms. An appropriate design of these interpenetrating forms makes a landscape into a dwelling place, and we learn to inhabit the world.

# Conclusion

I began by asking why we have grown so estranged from nature. We recognize that our origins lie in a mythological world. In the mythological experience, humans are part of a whole unified entity and there is not yet an "I" separate and distinct from the "not-I." In the mythological world, humans are connected with the things around them through forces that are not readily specifiable. Until their expulsion from the Garden of Eden, humans had been insiders; for them, there had not yet been an "outside" as we know it. The Fall inaugurated and set in motion self-awareness. Human beings learned to discover the "I" as separate and distinct from the "not-I." They learned to challenge their state of being linked to nature, to raise themselves above the state of nature.

As soon as people relinquish nature, nature enters the state of the wilderness – wilderness being nature in the absence of human beings. Wilderness, spreading beyond the garden after the Fall, threatens us in our very existence. It is therefore to be replaced with a structure governed by the *logos*, our own thinking transformed into reality – rational spaces. We in our state of profanity are convinced that we alone make ourselves and our world. We consider the mind, as the instrument of reason, to be separate from sensuous experience. The knowledge of nature that we acquire is founded in reason, and especially in mathematical reason. Mathematical reason creates space that is by itself, and it contains bodies whose position is not determined by the nature of the bodies themselves. We need to know what *is* without any subjective contributions. In order to study things objectively, we mistrust our own experience and we construct in thought conceptually the one objective world that *is*, irrespective of whether we are present or not.

Today, we customarily adopt two attitudes vis-à-vis nature, attitudes that are at the same time related and contradictory. The position I have just described is that of the impartial observer who intends to build a physical theory of nature to serve in the construction of rational spaces. The rational spaces separate us and protect us from nature in the state of the wilderness. The other attitude is that of excluding ourselves as human beings from nature in order to conserve the state of nature so that we may study it in its original condition. Stepping outside as impartial watcher in order to gain the knowledge of nature as the objective world contrasts with finding ourselves inevitably amidst the things in our life-world. Constructing the physical theory of nature on the basis of acknowledging solely objects is very different from experiencing aspects of nature in our everyday life. The knowledge of the objective world was meant to serve us in the betterment of our life-world, but we have overlooked the fact that it is our everyday experience, our life-world, that gives us access to what *is*.

In our life-world we are present as path subjects. As hikers, for instance, our path may be in the form of a gestalt course, shaped entirely by the gestalten all around us. Through this kind of movement we are part of a whole spatial organization. Otto Friedrich Bollnow points out the importance of hiking because, he says, the hiker is no longer separate from the landscape. The landscape makes the hiker into a part of itself and the components of the landscape appear to the hiker in their own light.[1]

In our life-world we have space and are connected to the things, and we also make space by building locations. It is the body subject, an entity in which mind and body are joined, that forms the instrument of our experience. A body subject is not in space; rather, it exists in a place, and a place is a significant centre of our immediate experience of the world. A place invites us to enter and assists us in orientating ourselves in relation to the world. Furthermore, places may induce us to perform a pleasurable gestalt course and thereby open ourselves to the world: "Yet friends, the church bell is just telling us the sixth hour – we want to walk to the 'Small Lucin' before supper. There we see the leafy crowns of the trees which give the water a deep green colour, a curious phenomenon which one cannot see often enough."[2] But a place is not a natural structure; it is a cultural form. Only through living in places are we able to build a world. It is our human condition that we have to constitute a world. We are able to build a world by bringing aspects of what *is* into our presence. What *is* is not primarily something by itself; rather, it constitutes itself in our experience. It comes into "being" and it takes its characteristic shape only when our intentions and aspects of the world meet and form a relationship.

We may bring aspects of nature into our presence by shaping them into "nature as landscape." A landscape represents a physiognomy and it is characteristic of any physiognomy that it has an outer aspect – how something appears to us when it is present – and this outer "face" allows us to grasp the inner nature of it. In order to understand a physiognomy, it is necessary to acquire the skill of apprehending what is appearing and to interpret it. Inherent to interpretation is, however, the possibility that we may interpret erroneously. Peirce, we recall, distinguishes between the "dynamic interpretant" and the "final interpretant." The dynamic interpretant is the reaction brought about in the interpreter by a sign. The final interpretant is the interpretation which the interpreter is supposed to attain if the sign is fully considered.

Nature has become present together with humankind. Both are linked in an extraordinary manner in becoming what they not yet are. As persons, we are in the process of becoming, and the object towards which we give our attention is not yet what it is meant to be. We are not present in an objective world; we are present in a situation. As noted earlier, Jung considers that one of the principal functions of the soul is to enable us to anticipate the possibilities that lie within a situation.

In a situation, we are present together with the things and we have to form a relationship with them. Building a relationship depends on the mode in which we stand to the things and how we act towards them. Moshé Feldenkrais, a student of human body movement and behaviour, points out that if we do not know what we actually do, we are incapable of doing what we wish to do.[3] Once we learn to act in various different ways, we may act in relation to things habitually by building up a structure of experience based on practical understanding. Instead of building a structure of experience based on theoretical knowing within one objective world, we need to build a structure of experience based on practical understanding within the context of our life-world.

We need to design our environment so that it becomes our place of habitation. This implies a place where we adopt the appropriate habits. The world we create arises from our habits, how we inhabit. We need to free ourselves by creating a range of habits of action and an environment that enhances the quality of our habits of action. How are we to judge the quality of our habits? The quality of our habits may be recognized by the extent to which our acts produce what cannot arise by itself in the manner in which it should.

If we have come to be present together with nature, why have we become so estranged from nature? It is because we have forgotten about "the mute things which are all around us" and instead have "our gaze fixed on remote enterprises" and are "embarked upon the conquest of distant schematic cities," as Ortéga expresses it. Our own admission of profound weakness, such as ignorance and affectation, requires us to limit the expression of our wants. We must commit ourselves to the cultural sublimation of our desires and the enhancement rather than disfigurement of our fellow human beings and natural beings, and our common dwelling place, the earth.

# Notes

INTRODUCTION

1 Rainer Maria Rilke, "Worpswede," in *Rainer Maria Rilke: Ausgewählte Werke* (Wiesbaden: Insel Verlag, 1951), 2:228–9 (my translation).
2 *Deutsches Wörterbuch*, 1889, s.v. "Naturwissenschaft" (my translation).
3 Jakob von Uexküll, *Niegeschaute Welten* (Berlin: S. Fischer, 1936), 11 (my translation).
4 Jakob von Uexküll, *Theoretische Biologie* (1928; reprint, Frankfurt am Main: Suhrkamp, 1973).
5 Edmund Husserl, *Ideas: General Introduction to Pure Phenomenology*, trans. W.R. Boyce Gibson (London: George Allen and Unwin, 1931), 241–51.
6 José Ortega y Gasset, *Meditations on Quixote*, trans. Evelyn Rugg and Diego Marín (New York: W.W. Norton, 1963), 43.

CHAPTER ONE

1 Conrad Gesner, *Thierbuch* (Zürich: Froschower, 1563).
2 Theodor Ballauff, *Die Wissenschaft vom Leben: Eine Geschichte der Biologie*, vol. 1, *Vom Altertum bis zur Romantik* (Freiburg: Karl Alber, 1954), 136.
3 Wilfried Blunt, *The Complete Naturalist* (London: Collins, 1971), 115.
4 Charles Robert Darwin, *Autobiography and Selected Letters*, ed. Francis Darwin (New York: Dover, 1958), 44.

5  Alexander von Humboldt und Aimé Bonpland, *Ideen zu einer Geographie der Pflanzen nebst einem Gemälde der Tropenländer* (Tübingen: F.G. Cotta, 1807).

6  Martin Heidegger, "The Origin of the Work of Art," in *Martin Heidegger: Basic Writings,* ed. David Farrell Krell (New York: Harper and Row, 1977), 153–60.

7  Georg Schmidt, "Naturalismus und Realismus (1959)," in Schmidt, *Umgang mit Kunst: Ausgewählte Schriften 1940–1963* (Basel: Verein der Freunde des Kunstmuseums, 1976), 27–36.

8  Wilhelm Bölsche, *Die naturwissenschaftlichen Grundlagen der Poesie,* ed. Johannes J. Braakenburg (Munich: Deutscher Taschenbuch-Verlag, 1976), 7–8 (my translation).

9  Gertrud Lenzer, ed., *Auguste Comte and Positivism: The Essential Writings* (New York: Harper and Row, 1975), 71–86.

10  Paul Arthur Schilpp, ed., *Albert Einstein: Philosopher Scientist* (New York: Tudor Publishing, 1951), 669.

11  Edward Batschelet, *Introduction to Mathematics for Life Scientists* (New York: Springer-Verlag, 1971), 53–7.

12  Felix Kaufmann, "Cassirer's Theory of Scientific Knowledge," in *The Philosophy of Ernst Cassirer,* ed. Paul Arthur Schilpp (LaSalle, Ill.: Open Court Publishing, 1973), 183–213.

13  F.S.C. Northrop, "The Method and Theories of Physical Science in their Bearing on Biological Organization," *Growth* supplement (1940): 127–54.

CHAPTER TWO

1  Mircea Eliade, *The Sacred and the Profane: The Nature of Religion,* trans. William R. Trask (New York: Harcourt, Brace and World, 1959), 11.

2  Rudolf Otto, *The Idea of the Holy: An Inquiry into the Non-Rational Factor in the Idea of the Divine and its Relation to the Rational,* trans. John W. Harvey, 2d ed. (London: Oxford University Press, 1957), 5–11.

3  Lucien Levy-Bruhl, *How Natives Think,* trans. Lilian A. Clare (New York: Washington Square Press, 1966), 3.

4  Ibid., 13.

5  Ibid., 25.

6  Gen. 2:8–9.

7  Gen. 2:15–17.

8  Gen. 3:7.

9  Heinz Westman, "Die Erlösungsidee im Judentum," in *Gestalt der*

*Erlösungsidee im Judentum und im Protestantismus, Eranos Jahrbuch 4 – 1936*, ed. Rudolf Ritsema (Ascona: Eranos Stiftung, 1986), 36.

10 Heinz Heimsoeth, *Die sechs grossen Themen der abendländischen Metaphysik und der Ausgang des Mittelalters*, 6th ed. (Darmstadt: Wissenschaftliche Buchgesellschaft, 1974), 90–130.

11 Francesco Petrarcha, "The Ascent of Mont Ventoux," trans. Hans Nachod, in *The Renaissance Philosophy of Man*, ed. Ernst Cassirer et al. (Chicago: The University Press, 1948), 36–46.

12 Ibid., 44.

13 Roderick Nash, *Wilderness and the American Mind* (New Haven: Yale University Press, 1973).

14 Oto Bihalji-Merin, *Masters of Naive Art*, trans. Russell M. Stockman (New York: McGraw-Hill, n.d.), 69.

15 Ezek. 29:5.

16 Otto Friedrich Bollnow, *Mensch und Raum*, 4th ed. (Stuttgart: W. Kohlhammer, 1980), 31–7.

### CHAPTER THREE

1 Heinrich von Kleist, "Über das Marionettentheater," in *Heinrich von Kleist: Werke in einem Band* (Munich: Carl Hanser Verlag, 1960), 802–7.

2 Ibid., 807 (my translation).

3 *Descartes: Philosophical Writings*, selected and trans. Norman Kemp Smith (New York: Modern Library, 1958), 299.

4 Hans-Wilhelm Koepke, *Die Lebensformen* (Krefeld: Goecke & Evers, 1973), 158–75.

5 Ibid., 287.

6 Ibid., 289–90.

7 Ibid., 298–9.

8 Gustav Theodor Fechner, *Elements of Psychophysics*, vol. 1, trans. Helmut E. Adler, ed. Davis H. Howes and Edwin G. Boring (New York: Holt, Rinehart and Winston, 1966).

9 Norbert Wiener, *Cybernetics – or Control and Communication in the Animal and the Machine* (New York: John Wiley, 1948).

### CHAPTER FOUR

1 Eduard Spranger, "Zur Theorie des Verstehens und zur geisteswissen-schaftlichen Psychologie (1918)," in *Grundlagen der Geisteswissenschaften*, ed. Hans Walter Bahr (Tübingen: Max Niemeyer Verlag, 1980), 1–42.

2 Ibid., 15.

3 Wilhelm Dilthey, "Weltanschauung und Analyse des Menschen seit Renaissance und Reformation," in *Wilhelm Dilthey: Gesammelte Schriften* (Stuttgart: B.G. Teubner, 1977), 2:1–15.

4 Friedrich Nietzsche, *The Joyful Wisdom*, trans. Thomas Common (New York: Russell and Russell, 1964), 167.

5 Julian Jaynes, *The Origin of Consciousness in the Breakdown of the Bicameral Mind* (Boston: Houghton Mifflin, 1982).

6 Nietzsche, *Joyful Wisdom*, 276.

7 Martin Heidegger, "The Origin of the Work of Art," in *Martin Heidegger: Basic Writings*, ed. David Farrell Krell (New York: Harper and Row, 1977), 162.

8 Hermann von Helmholtz, "On the Interaction of Natural Forces," trans. Professor Tyndall, in *Popular Lectures on Scientific Subjects by H. Helmholtz*, trans. E. Atkinson (London: Longmans, Green, 1884), 137–71.

9 *Webster's Collegiate Dictionary*, 5th ed., s.v. "need."

10 Martin Heidegger, "Modern Science, Metaphysics, and Mathematics," in *Martin Heidegger: Basic Writings*, 260–5.

11 Martin Heidegger, "The Question Concerning Technology," in *Martin Heidegger: Basic Writings*, 295–6.

12 Ibid., 287.

13 Ibid., 295.

14 Ibid., 298.

15 Ibid., 298, 297, 300, 299.

16 Ibid., 305.

17 Ibid., 307–9.

CHAPTER FIVE

1 Martin Heidegger, "The Origin of the Work of Art," in *Martin Heidegger: Basic Writings*, ed. David Farrell Krell (New York: Harper and Row, 1977), 149–87.

2 K.J. Narr, "Beiträge der Urgeschichte zur Kenntnis der Menschennatur," in *Neue Anthropologie: Kulturanthropologie*, ed. Hans-Georg Gadamer and Paul Vogler (Stuttgart: Georg Thieme Verlag, 1973), 4.

3 A. Thaer, quoted by Alfred Barthelmess, *Landschaft: Lebensraum des Menschen* (Freiburg, Breisgau: Alber Verlag, 1987), 215 (my translation).

4 Ludwig Klages, "Mensch und Erde," in Klages, *Mensch und Erde: Gesammelte Abhandlungen* (Stuttgart: Kröner, 1973), 1–25.

5 Norman Myers, *The Sinking Ark: A New Look at the Problem of Disappearing Species* (Oxford: Pergamon Press, 1979).

6 G. Tyler Miller, Jr, *Living in the Environment: An Introduction to Environmental Science,* 4th ed. (Belmont, Calif.: Wadsworth Publishing, 1985), 4–13.

7 Ibid., E80–5.

8 Friedrich Kaulbach, *Einführung in die Philosophie des Handelns* (Darmstadt: Wissenschaftliche Buchgesellschaft, 1986).

9 Ibid., 56–63.

10 Timothy O'Riordan, *Environmentalism* (London: Pion, 1976), 1–2.

11 Rudolf Kassner, "Die Moral der Musik," in *Rudolf Kassner: Sämtliche Werke,* ed. Ernst Zinn (Pfullingen: Günther Neske, 1969), 1:491–755.

12 Neil Evernden, *The Natural Alien: Humankind and Environment* (Toronto: University of Toronto Press, 1985).

13 Erwin Straus, "Norm and Pathology of I-World Relations," in *Selected Papers of Erwin W. Straus: Phenomenological Psychology,* trans. Erling Eng (New York: Basic Books, 1966), 255–76.

14 Walter Falk, "Impressionismus und Expressionismus," in *Expressionismus als Literatur,* ed. Wolfgang Rothe (Bern: Francke Verlag, 1969), 69–86.

15 Georg Heym, "Der Gott der Stadt," in *Lyrik des expressionistischen Jahrzehnts,* introd. Gottfried Benn (Wiesbaden: Limes Verlag, 1974), 66.

16 Thure von Uexküll and Wolfgang Wesiak, "Psychosomatische Medizin und das Problem einer Theorie der Heilkunde," in *Lehrbuch der psychosomatischen Medizin,* ed. Thure von Uexküll (Munich: Urban & Schwarzenberg, 1981), 16.

17 Cecile Ernst, "Kontroverse Schizophrenie," *Neue Zürcher Zeitung* 221 (1981): 37 (my translation).

18 Eugène Minkowski, *Lived Time: Phenomenological and Psychopathological Studies,* trans. Nancy Metzel (Evanston, Ill.: Northwestern University Press, 1970).

19 R.D. Laing and A. Esterson, *Sanity, Madness and the Family* (Harmondsworth: Penguin, 1980), 49.

20 Ibid., 74.

21 Minkowski, *Lived Time,* 416–22.

22 Ibid., 406.

CHAPTER SIX

1 James J. Gibson, "The Theory of Affordances," in *Perceiving, Acting, and Knowing: Toward an Ecological Psychology*, ed. Robert Shaw and John Bransford (Hillsdale, N.J.: Lawrence Erlbaum Assoc., 1977), 67–82.

2 Jakob von Uexküll, *Theoretische Biologie* (1928; reprint, Frankfurt am Main: Suhrkamp, 1973).

3 Ibid., 158.

4 Jakob von Uexküll, *Niegeschaute Welten* (Berlin: S. Fischer, 1936), 14–15 (my translation).

5 Jakob von Uexküll and Georg Kriszat, *Streifzüge durch die Umwelten von Tieren und Menschen/Bedeutungslehre* (Stuttgart: S. Fischer, 1970), 126 (my translation).

6 A. Frey-Wyssling, *Stoffwechsel der Pflanzen* (Zürich: Büchergilde Gutenberg, 1949), 229–31.

7 Brian Hocking, "Ant-Plant Mutualism: Evolution and Energy," in *Coevolution in Animals and Plants*, ed. Lawrence E. Gilbert and Peter H. Raven (Austin: University of Texas Press, 1975), 78–90.

8 Ibid., 82.

9 Charles Hartshorne and Paul Weiss, eds., *Collected Papers of Charles Sanders Peirce* 8 vols. (Cambridge, Mass.: Harvard University Press, 1932–1958), 4:422.

10 Ibid., 5:330.

11 Charles Morris, "Signs and the Act," in *Charles Morris: Writings on the General Theory of Signs*, ed. Thomas A. Sebeok (The Hague: Mouton, 1971), 401–14.

12 *The Oxford English Dictionary*, 1933, s.v. "symptomatology."

13 Ibid., s.v. "symptom."

14 Charles S. Peirce, "Letters to Lady Welby," in *Values in a Universe of Chance: Selected Writings of Charles S. Peirce (1839–1914)*, ed. Philip P. Wiener (Garden City, N.Y.: Doubleday, 1958), 413–14.

15 Gustav Vriesen, *August Macke* (Stuttgart: W. Kohlhammer Verlag, 1957), 260–1 (my translation).

16 Hartshorne and Weiss, *Collected Papers of Charles Sanders Peirce*, 1:227–80.

17 Peirce, "Letters to Lady Welby," 388.

18 von Uexküll and Kriszat, *Streifzüge*, 114–17.

19 Hartshorne and Weiss, *Collected Papers of Charles Sanders Peirce*, 5:332.

20 Ibid., 2:143.

21  Ibid.
22  Ibid., 8:92.

CHAPTER SEVEN

1  Martin Buber, "Distance and Relation," in Buber, *The Knowledge of Man: Selected Essays,* ed. Maurice Friedman, trans. Maurice Friedman and Ronald Gregor Smith (Atlantic Highlands, N.J.: Humanities Press International, 1988), 50.
2  Helmuth Plessner, "Die Stufen des Organischen und der Mensch," in *Helmuth Plessner: Gesammelte Schriften,* vol. 4, ed. Günter Dux, Odo Marquard, and Elisabeth Ströker (Frankfurt am Main: Suhrkamp Verlag, 1981).
3  Ibid., 303–11.
4  José Ortega y Gasset, "Vitalität, Seele, Geist," in *José Ortega y Gasset: Gesammelte Werke* (Stuttgart: Deutsche Verlags-Anstalt, 1978), 1:317–50.
5  Maurice Merleau-Ponty, *Phenomenology of Perception,* trans. Colin Smith (New York: Humanities Press, 1962).
6  José Ortega y Gasset, *Meditations on Hunting,* trans. Howard B. Wescott (New York: Charles Scribner's Sons, 1972), 90.
7  Ibid., 142.
8  Edmund Husserl, *The Crisis of European Sciences and Transcendental Phenomenology,* trans. David Carr (Evanston, Ill.: Northwestern University Press, 1970), 103–89.
9  Victor von Weizsäcker, *Gestalt und Zeit,* 2d ed. (Göttingen: Vanden Hoek & Ruprecht, 1960), 15–16.
10  Victor von Weizsäcker, *Der Gestaltkreis,* 4th ed. (Stuttgart: Georg Thieme, 1950).
11  von Weizsäcker, *Gestalt und Zeit,* 45.
12  Ibid., 16.
13  von Weizsäcker, *Der Gestaltkreis,* 149–51.
14  Hermann Schmitz, *Der Leib im Spiegel der Kunst* (Bonn: H. Bouvier, 1966), 5.
15  Ibid., 34.
16  Christian von Ehrenfels, "Über Gestaltqualitäten," in *Gestalthaftes Sehen,* ed. Ferdinand Weinhandl (Darmstadt: Wissenschaftliche Buchgesellschaft, 1960), 11–43.
17  Schmitz, *Der Leib im Spiegel der Kunst,* 23 (my translation).

CHAPTER EIGHT

1  James J. Gibson, *The Ecological Approach to Visual Perception* (Boston: Houghton Mifflin, 1979), 65–92.
2  Ibid., 73.
3  Ibid., 227–9.
4  Hermann Schmitz, *Der Leib im Spiegel der Kunst* (Bonn: H. Bouvier, 1966), 5.
5  Erwin W. Straus, "The Forms of Spatiality," in *Selected Papers of Erwin W. Straus: Phenomenological Psychology*, trans. Erling Eng (New York: Basic Books, 1966), 3–37.
6  Ibid., 11.
7  Kurt Rowland, *A History of the Modern Movement: Art Architecture Design*, (New York: Van Nostrand Reinhold, 1973), 2:92–3.
8  Anya Peterson Royce, *The Anthropology of Dance* (Bloomington: Indiana University Press, 1977), 199.
9  Straus, "Forms of Spatiality," 32.
10  Jan Mukarovsky, "The Place of the Aesthetic Function among the Other Functions," in *Structure, Sign and Function: Selected Essays by Jan Mukarovsky*, trans. and ed. John Burbank and Peter Steiner (New Haven, Conn.: Yale University Press, 1978), 39.
11  Eduard Spranger, *Lebensformen*, 9th ed. (Tübingen: Max Niemeyer Verlag, 1966).
12  David Seamon, "Body-Subject, Time-Space Routines, and Place-Ballets," in *The Human Experience of Space and Place*, ed. Anne Buttimer and David Seamon (London: Croom Helm, 1980), 157–8.
13  Ibid., 159.
14  Lewis Mumford, *Culture of Cities* (New York: Harcourt Brace Jovanovich, 1970), 51.

CHAPTER NINE

1  Helmuth Plessner, "Die Stufen des Organischen und der Mensch," in *Helmuth Plessner: Gesammelte Schriften* (Frankfurt am Main: Suhrkamp, 1981), 4:360–4.
2  John Locke, *An Essay Concerning Human Understanding*, collated and annotated by Alexander Campbell Fraser (New York: Dover, 1959), 1:166–82.
3  Ibid., 174.

4 Agnes Arber, *The Mind and the Eye* (Cambridge: The University Press, 1954), 125.

5 Susanne K. Langer, *Philosophy in a New Key* (New York: Mentor Books, 1942), 65–6.

6 Thomas A. Sebeok, *Perspectives in Zoosemiotics* (The Hague: Mouton, 1972).

7 Langer, *Philosophy in a New Key*, 75–9.

8 Konrad Lorenz, "Gestaltwahrnehmung als Quelle wissenschaftlicher Erkenntnis" (1959), in Lorenz, *Über tierisches und menschliches Verhalten*, 11th ed. (Munich: R. Piper Verlag, 1974), 1:255–300.

9 Arber, *The Mind and the Eye*.

10 Ibid., 117.

11 Rudolf Arnheim, *Visual Thinking* (Berkeley: University of California Press, 1989).

12 Ibid., 174.

13 Ibid., 117–88.

14 Johann Caspar Lavater, *Essays on Physiognomy: For the Promotion of the Knowledge and the Love of Mankind*, 3rd ed., trans. Thomas Holcroft (London: Blake, 1840).

15 Alexander von Humboldt, *Views of Nature*, trans. E.C. Otté and Henry G. Bohn (New York: Arno Press, 1975).

16 Michael Polanyi, *Personal Knowledge: Towards a Post-Critical Philosophy* (Chicago: University of Chicago Press, 1962).

17 Michael Polanyi, *The Tacit Dimension* (Garden City, N.Y.: Doubleday, 1966), 10.

18 Michael Polanyi and Harry Prosch, *Meaning* (Chicago: University of Chicago Press, 1975), 33.

19 Polanyi, *Tacit Dimension*, 10.

20 Ibid., 35–6.

21 Polanyi and Prosch, *Meaning*, 36.

22 Polanyi, *Tacit Dimension*, 40–2.

23 Ibid., 41.

CHAPTER TEN

1 K.J. Narr, "Beiträge der Urgeschichte zur Kenntnis der Menschennatur," in *Neue Anthropologie: Kulturanthropologie*, ed. Hans-Georg Gadamer and Paul Vogler (Stuttgart: Georg Thieme Verlag, 1973), 4.

2   Timothy O'Riordan, *Environmentalism* (London: Pion, 1976), 1–2.

3   Georg Simmel, "Zur Philosophie der Kultur," in Simmel, *Philosophische Kultur: Gesammelte Essais* (Berlin: Verlag Klaus Wagenbach, 1983), 183.

4   Jolande Jacobi, *The Psychology of C.G. Jung*, trans. Ralph Manheim (New Haven: Yale University Press, 1974).

5   Carl Gustav Jung, "Spirit and Life," in *C.G. Jung: Collected Works* (Princeton: University Press, 1969), 8:329.

6   Martin Heidegger, "The Word of Nietzsche 'God is Dead,'" in Heidegger, *The Question Concerning Technology and Other Essays*, trans. William Lovitt (New York: Harper and Row, 1977), 71.

7   Rudolf Kassner, "Die Moral der Musik," in *Rudolf Kassner: Sämtliche Werke*, ed. Ernst Zinn (Pfullingen: Günter Neske, 1969), 1:543–4.

8   José Ortega y Gasset, *Meditations on Quixote*, trans. Evelyn Rugg and Diego Marín (New York: W.W. Norton, 1963), 45.

9   Ibid., 59–60.

10   Ibid., 41.

11   Martin Buber, "Distance and Relation," in Buber, *The Knowledge of Man: Selected Essays*, ed. Maurice Friedman, trans. Maurice Friedman and Ronald Gregor Smith (Atlantic Highlands, N.J.: Humanities Press International, 1988), 56.

12   Rainer Marie Rilke, "Von der Landschaft," in *Rainer Maria Rilke: Ausgewählte Werke* (Wiesbaden: Insel Verlag, 1951), 2:221 (my translation).

13   Ibid., 222 (my translation).

14   Ibid., 223–4 (my translation).

15   Ibid., 225 (my translation).

16   Nick Johnson, "Natural Grace in the Work of Jesus Soto," *Arts Canada* 35 (1978): 11.

17   Klaus Lankheit, *Franz Marc: Sein Leben und seine Kunst* (Cologne: DuMont, 1976), 44–5.

18   Klaus Lankheit, ed., *Franz Marc: Schriften* (Cologne: DuMont, 1978), 112 (my translation).

19   Kurt Rowland, *A History of the Modern Movement: Art Architecture Design* (New York: Van Nostrand Reinhold, 1973), 2:100.

20   Martin Buber, "Distance and Relation," 56.

CHAPTER ELEVEN

1   *Deutsches Wörterbuch*, 1885, s.v. "Landschaft" (my translation).

2   J. Brian Bird, *The Natural Landscapes of Canada: A Study in Regional Earth Science* (Toronto: Wiley Publications, 1972).

3   Ibid., 71.

4   Carl Troll, "Klima und Pflanzenkleid der Erde in dreidimensionaler Sicht" (1960), in *Pflanzengeographie*, ed. Wilhelm Lauer and Hans-Jürgen Klink (Darmstadt: Wissenschaftliche Buchgesellschaft, 1978), 355–400.

5   Jay Appleton, *The Experience of Landscape* (London: John Wiley and Sons, 1975).

6   Fred Fischer, *Der animale Weg* (Zürich: Artemis, 1972), 9–50.

7   Carl Gustav Jung, *Memories, Dreams, Reflections*, ed. Aniela Jaffe, trans. Richard and Clara Winston (New York: Vintage Books, 1963), 255.

8   E. Relph, *Place and Placelessness* (London: Pion, 1976).

9   August Nitschke, *Kunst und Verhalten: Analoge Konfigurationen* (Stuttgart-Bad Canstatt: Fromman-Holzboog, 1975), 14.

10  Wolf Jobst Siedler, "Welt ohne Schatten," in Ernst Jünger and Wolf Jobst Siedler, *Bäume* (Frankfurt am Main: Ullstein, 1976), 89.

11  Jan Mukarovsky, "The Place of the Aesthetic Function among the Other Functions," in *Structure, Sign and Function: Selected Essays by Jan Mukarovsky*, trans. and ed. John Burbank and Peter Steiner (New Haven: Yale University Press, 1978), 39.

12  F.-W. von Hermann, *Heideggers Philosophie der Kunst* (Frankfurt am Main: Vittorio Klostermann, 1980), 32–4.

13  Carl Gustav Jung, "Psychological Types," in *C.G. Jung: Collected Works*, ed. Sir Herbert Read et al. (Princeton: University Press, 1971), 6:467–8.

14  Neil Evernden, "Beauty and Nothingness," *Wildflower* 3 (1987): 22–5.

15  Ibid., 25.

16  Georg Simmel, "Zur Philosophie der Kultur," in Simmel, *Philosophische Kultur: Gesammelte Essais* (Berlin: Verlag Klaus Wagenbach, 1983), 183.

17  Christian Norberg-Schulz, *Genius Loci: Towards a Phenomenology of Architecture* (New York: Rizzoli, 1979), 19–21.

CHAPTER TWELVE

1   Ernst Cassirer, *An Essay on Man* (New Haven: Yale University Press, 1962), 33.

2   E. Relph, *Place and Placelessness* (London: Pion, 1976).

3   José Ortega y Gasset, *Meditations on Quixote*, trans. Evelyn Rugg and Diego Marín (New York: W.W. Norton, 1963), 45.

4  Ibid., 41–6.
5  Gottfried Benn, "Goethe und die Naturwissenschaften," in *Gottfried Benn: Das Hauptwerk*, ed. Marguerite Schüter (Wiesbaden: Limes Verlag, 1980), 2:95.
6  Martin Heidegger, "Building Dwelling Thinking," in *Martin Heidegger: Basic Writings*, ed. David Farrell Krell (New York: Harper and Row, 1977), 333.
7  Martin Buber, "Distance and Relation," in Buber, *The Knowledge of Man: Selected Essays*, ed. Maurice Friedman, trans. Maurice Friedman and Ronald Gregor Smith (Atlantic Highlands, N.J.: Humanities Press International, 1988), 56.
8  Theodore Reff, "Painting and Theory in the Final Decade," in *Cézanne: The Late Work* (New York: Museum of Modern Art, 1977), 13–53.
9  Martin Heidegger, "The Origin of the Work of Art," in *Martin Heidegger: Basic Writings*, 143–87.
10  E. Relph, *Place and Placelessness*, 49.
11  James R. Lawler, ed., *Paul Valéry: An Anthology* (Princeton: Princeton University Press, 1977), 71, 79, 83–4.
12  Christopher Alexander, *Notes on the Synthesis of Form* (Cambridge, Mass.: Harvard University Press, 1974).
13  Emil Nolde, "Urvölkerkunst," in Nolde, *Mein Leben* (Cologne: DuMont, 1976), 201–2 (my translation).
14  Alexander, *Notes*, 84–94.
15  Charles S. Peirce, "Letters to Lady Welby," in *Values in a Universe of Chance: Selected Writings of Charles S. Peirce (1839–1914)*, ed. Philip P. Wiener (Garden City, N.Y.: Doubleday, 1958), 383.
16  Carlo Huber, *Otto Meyer-Amden* (Wabern: Buchler Verlag, 1968), 50.
17  Georg Trakl, "Ein Winterabend," in *Georg Trakl: Die Dichtungen* (Salzburg: Otto Müller Verlag, 1978), 120.
18  Edward Trier, *Figur und Raum: Die Skulptur des 20. Jahrhunderts* (Berlin: Gebr. Mann Verlag, 1960), 9–12.
19  Erwin W. Straus, "The Forms of Spatiality," in *Selected Papers of Erwin W. Straus; Phenomenological Psychology*, trans. Erling Eng (New York: Basic Books, 1966), 11.
20  Nick Johnson, "Natural Grace in the Work of Jesus Soto," *Arts Canada* 35 (1978): 11.
21  Ibid., 8.
22  Martin Heidegger, "Building Dwelling Thinking," in *Martin Heidegger: Basic Writings*, 327.

23 Martin Heidegger, *Sein und Zeit* (Tübingen: Neomarius Verlag, 1949), 53.

24 Martin Heidegger, "The Thing," in Heidegger, *Poetry, Language, Thought*, trans. Albert Hofstadter (New York: Harper and Row, 1975), 163–86.

25 Franz Marc, *Briefe 1914–1916 Aus dem Felde* (Berlin: Rembrandt-Verlag, 1959), 8–9.

CONCLUSION

1 Otto Friedrich Bollnow, *Mensch und Raum*, 4th ed. (Stuttgart: Kohlhammer, 1980).

2 Fritz Meidner, "Kleinod des Landes," in *Brevier von Mecklenburg* (Hamburg: Hoffmann und Campe, 1958), 12 (my translation).

3 Moshé Feldenkrais, *The Elusive Obvious* (Cupertino: Meta Publications, 1981).

# Index